*Stefan Florczyk*

**Robot Vision**

*Stefan Florczyk*

# Robot Vision

## Video-based Indoor Exploration with Autonomous and Mobile Robots

WILEY-VCH Verlag GmbH & Co. KGaA

**Author**

***Dr. Stefan Florczyk***
Munich University of Technology
Institute for Computer Science
florczyk@in.tum.de

**Cover Picture**
They'll Be More Independent, Smarter and More Responsive




**Library of Congress Card No.:**
applied for

**British Library Cataloguing-in-Publication Data**
A catalogue record for this book is available from the British Library.

**Bibliographic information published by Die Deutsche Bibliothek**
Die Deutsche Bibliothek lists this publication in the Deutsche Nationalbibliografie; detailed bibliographic data is available in the Internet at <http://dnb.ddb.de>.



Printed in the Federal Republic of Germany.

Printed on acid-free paper.

**Typesetting** Kühn & Weyh, Satz und Medien, Freiburg
**Printing** betz-druck GmbH, Darmstadt
**Bookbinding** Litges & Dopf Buchbinderei GmbH, Heppenheim

**ISBN** 3-527-40544-5

*Dedicated to my parents*

# Contents



*Robot Vision: Video-based Indoor Exploration with Autonomous and Mobile Robots.* Stefan Florczyk

ISBN: 3-527-40544-5

## List of Figures



*Robot Vision: Video-based Indoor Exploration with Autonomous and Mobile Robots.* Stefan Florczyk

ISBN: 3-527-40544-5

## Symbols and Abbreviations

| | |
|---|---|
| $A$ | Region |
| $AO$ | Association operators |
| $AW$ | Window |
| $B$ | Blue channel |
| $BC_j$ | Plane image contour $j$ |
| $BE$ | Base elements |
| $C$ | CCD array |
| $CE$ | Complex elements |
| $CO$ | Contour |
| $D$ | Projection matrix |
| $E$ | Essential matrix |
| $\mathrm{E}[x]$ | Expected value for random vector $x$ |
| $\mathrm{E}(A)$ | Erosion of a pixel set $A$ |
| $F$ | Focal point |
| $\mathrm{F}\{\mathrm{f}(\mathrm{x})\}$ | Fourier transform for function $\mathrm{f}(\mathrm{x})$ |
| $G$ | Green channel |
| $\mathrm{G}(f_x)$ | Fourier transform for Gabor filter |
| $H$ | The horizontal direction vector of the $Fz_CHL$ projection equation |
| $H(t)$ | Observation matrix |
| I | Image |
| $J$ | Jacobi matrix |
| $K$ | Matrix |
| $L_i$ | The Jacobi matrix of the observation equation at point in time $i$ |
| L | Left-hand side |
| $L$ | The vertical direction vector of the $Fz_CHL$ projection equation |
| $M$ | Rotation matrix |
| $N$ | Unit matrix |
| $O$ | Origin |
| $P$ | Covariance matrix |
| $P_k^+$, $P_{k+1}^-$ | The update and prediction of the covariance matrix, respectively |
| $Q$ | Index set |
| $Q_i$ | The covariance matrix of process noise at point in time $i$ |

*Robot Vision: Video-based Indoor Exploration with Autonomous and Mobile Robots.* Stefan Florczyk

ISBN: 3-527-40544-5

| | |
|---|---|
| $R$ | Red channel |
| R | Right-hand side |
| $S$ | Structuring element |
| $SE$ | Set |
| $T$ | Rotation |
| Tf(x) | Vectorial transformation function |
| U | Disparity |
| $U$ | Color channel |
| $V$ | Color channel |
| $W$ | Skew symmetric matrix |
| W($A$) | Function that calculates the width of a region $A$ |
| $X$ | Point in three-dimensional space |
| $Y$ | Brightness |
| $Z$ | Diagonal matrix |
| $Z_i$ | Motion matrix at point in time $i$ |
| $a$ | Length |
| $ao$ | Association operator |
| $b$ | Focal length |
| $c$ | Constant |
| $d$ | Distance |
| $d_i$ | The binormal vector of a polygon with index $i$ |
| $dr_{\mathrm{max}}$ | Maximal recursion depth by the cluster search |
| $e_x^{\mathrm{C}}, e_y^{\mathrm{C}}, e_z^{\mathrm{C}}$ | The column vectors of transformation matrix $\mathrm{H}_{\mathrm{BC}_j,\mathrm{C}}$ in camera coordinates |
| $e$ | Epipole |
| $f$ | Result |
| $f_x$ | Frequency measured at $X$-axis |
| $f_{x\mathrm{m}}$ | Middle frequency measured at $X$-axis |
| $g$ | Epipolar line |
| $g_x$ | The size of a pixel in $X$ direction |
| $g_y$ | The size of a pixel in $Y$ direction |
| g(x) | One-dimensional Gabor filter |
| $h_i$ | The number of bisections per axis $i$ in the transformation space |
| $h$ | Principal point offset |
| i | Imaginary number |
| $ip$ | Vectorial input parameter |
| $l$ | Location |
| $lc$ | Contour length |
| $lg$ | The local goodness of single element $i$ in the transformation space |
| $m$ | Node |
| $n$ | Normal |
| $nf$ | Norm factor |
| $o$ | Octave |
| $of$ | Offset |
| $p$ | Pixel |

| | |
|---|---|
| $q(t)$ | State vector |
| $\tilde{q}(t)$ | Linear state valuer |
| $r(t)$ | Random vector |
| $r$ | Edge |
| $s$ | Spectrum factor |
| $se$ | Element in set $SE$ |
| $t$ | Time in general |
| $u$ | Projection |
| $v$ | Variable |
| $(x, y)_{\mathrm{A}}$ | Two-dimensional image affine coordinates |
| $(x, y)_{\mathrm{I}}$ | Two-dimensional image Euclidean coordinates |
| $(x, y)_{\mathrm{S}}$ | Two-dimensional sensor coordinates |
| $(x, y, z)_{\mathrm{W}}$ | Three-dimensional world coordinates |
| $(x, y, z)^{-}_{\hat{\mathrm{W}}_{k+1}}$ | The update of three-dimensional world coordinates at point in time $k+1$ |
| $(x, y, z)^{+}_{\hat{\mathrm{W}}_{k}}$ | The prediction of three-dimensional world coordinates at point in time $k$ |
| $A$ | The camera's line of sight |
| $B$ | Fundamental matrix |
| $\mathrm{B}(A)$ | Function that calculates the area of a region $A$ |
| $\mathbf{X}_{\mathrm{A}}$ | Image affine coordinate system with axes $X_{\mathrm{A}}$, $Y_{\mathrm{A}}$, and $Z_{\mathrm{A}}$ |
| $\mathbf{X}_{\mathrm{BC}_j}$ | Contour-centered coordinate system |
| $\mathbf{X}_{\mathrm{C}}$ | Camera Euclidian coordinate system with axes $X_{\mathrm{C}}$, $Y_{\mathrm{C}}$, and $Z_{\mathrm{C}}$ |
| $\mathbf{X}_{\mathrm{I}}$ | Image Euclidean coordinate system with axes $X_{\mathrm{I}}$, $Y_{\mathrm{I}}$, and $Z_{\mathrm{I}}$ |
| $\mathbf{X}_{\mathrm{M}}$ | Robot coordinate system |
| $\mathbf{X}_{\mathrm{S}}$ | Sensor coordinate system with axes $X_{\mathrm{S}}$, $Y_{\mathrm{S}}$, and $Z_{\mathrm{S}}$ |
| $\mathbf{X}_{\mathrm{T}}$ | Transformation space with axes $X_{\mathrm{T}}$, $Y_{\mathrm{T}}$, and $Z_{\mathrm{T}}$ |
| $\mathbf{X}_{\mathrm{W}}$ | World Euclidean coordinate system with axes $X_{\mathrm{W}}$, $Y_{\mathrm{W}}$, and $Z_{\mathrm{W}}$ |
| $\Delta$ | Difference |
| $H$ | Homogeneous matrix |
| $\mathrm{H}(A)$ | Function that calculates the height of a region $A$ |
| $I$ | Matrix |
| $K$ | Resolution power |
| $\Phi(t+1;t)$ | Transition matrix |
| $\Phi(t+1;t)^{*}$ | Estimated transition matrix |
| $\Gamma$ | Disparity gradient |
| $M$ | Matrix |
| $N$ | Matrix |
| $O$ | The vectorial output variable of a transformation |
| $P$ | Segment |
| $T$ | Threshold |
| $\Upsilon$ | $\Xi/K$ Quality |
| $\Omega$ | Tensor |
| $\Xi$ | Field of vision |
| $\Psi$ | Calibration matrix |

| | |
|---|---|
| $Z$ | Matrix |
| $\alpha$ | Rotation applied to $X$-axis |
| $\beta$ | Rotation applied to $Y$-axis |
| $\delta$ | Dilation |
| $\varepsilon$ | Error value |
| $\phi$ | Activation profile |
| $\varphi(x)$ | The local phase of a Gabor filter |
| $\gamma$ | Rotation applied to $Z$-axis |
| $\eta$ | Surface normal |
| $\iota$ | Intensity |
| $\kappa$ | Quality factor |
| $\lambda$ | Gabor wavelength |
| $\mu$ | Solution |
| $\mathrm{\mu}$ | Learning rate |
| $\nu$ | Vector |
| $o(t)$ | Observation vector |
| $\varpi$ | Probability |
| $\theta$ | Angle of rotation |
| $\vartheta$ | Object size |
| $\rho$ | Scale factor |
| $\sigma$ | Standard deviation |
| $\tau$ | Translation |
| $\upsilon$ | Neuron |
| $\omega$ | The element of a vector |
| $\omega(t)$ | Random vector |
| $\xi$ | Angle that is surrounded from two polygon areas |
| $\psi$ | Distortion coefficient |
| $\zeta$ | Balance |

# 1
# Introduction

The video-based exploration of interiors with autonomous and mobile service robots is a task that requires much programming effort. Additionally, the programming tasks differ in the necessary modules. Commands, which control the technical basis equipment, must consider the reality of the robot. These commands activate the breaks and the actuation. The parts are basically included in the delivery. Often a mobile robot additionally possesses sonar, ultrasonic, and cameras, which constitute the perception function of the robot. The programming of such a mobile robot is a very difficult task if no control software comes with the robot. First, the programmer must develop the necessary drivers. As a rule the manufacturer includes a software library into the scope of the supply. This enables programs in a high-level language like C++ to be created very comfortably to control most or all parts of the robot's basic equipment. Typically operators are available. The user can transfer values to the arguments whose domain depends on the device that is to be controlled, and the admitted measurement. Operators, which enable rotary motions, may take values in degrees or radians. The velocity can be adjusted with values, which are specified in meters per second or yards per second. Video cameras are sometimes also part of a mobile robot's basic equipment, but further software and/or hardware must be acquired generally. A frame grabber is required. This board digitizes the analog signal of the camera. The gained digital image can then be processed with an image-processing library. Such a library provides operators for the image processing that can also be included into a high-level program. If the camera comes with the robot, the manufacturer provides two radio sets if the computer that controls the robot is not physically compounded with the robot. One radio set is necessary to control the robot's basic equipment from a static computer. The second radio set transmits the analog camera signals to the frame grabber. Nowadays, robots are often equipped with a computer. In this case radio sets are not necessary, because data transfer between a robot's equipment and a computer can be directly conducted by the use of cables. Additionally, a camera head can be used that connects a camera with a robot and enables software-steered panning and tilting of the camera. Mobile service robots use often a laser that is, as a rule, not part of a robot. They are relatively expensive, but sometimes the robot-control software provided involves drivers for commercial lasers.

*Robot Vision: Video-based Indoor Exploration with Autonomous and Mobile Robots.* Stefan Florczyk

ISBN: 3-527-40544-5

The application areas for mobile service robots are manifold. For example, a realized application that guided people through a museum has been reported. The robot, named RHINO [1], was mainly based on a laser device, but many imaginable areas require the robot to be producible cheaply. Therefore, RHINO will not be further considered in this book. This book proposes robot-navigation software that uses only cheap off-the-shelf cameras to fulfill its tasks. Other imaginable applications could be postal delivery or a watchman in an office environment, service actions in a hospital or nursing home, and so forth. Several researchers are currently working on such applications, but a reliable application does not currently exist. Therefore, at this point a possible scenario for a mobile service robot that works autonomously will be illustrated.

Postal delivery is considered, as mentioned before. First, it should be said that such a task can not be realized with a robot that works exclusively with lasers, because the robot must be able to read. Otherwise it can not allocate letters to the particular addresses. Therefore, a camera is an essential device. If the mobile robot is to work autonomously, it is necessary that it knows the working environment. If the robot needs to be able to work in arbitrary environments, it is a problem if a human generates a necessary navigation map, which can be considered as a city map, offline. If the robot's environment changes, a new map must be created manually, which increases the operating costs. To avoid this, the robot must acquire the map autonomously. It must explore the environment before the operating phase starts. During the operating phase, the robot uses the created map to fulfill its tasks. Of course, its environment changes, and therefore it is necessary to update the map during operation. Some objects often change their positions like creatures; others remain rather permanently in the map. Desks are an example. The robot must also be able to detect unexpected obstacles, because collision avoidance must be executed. If letters are to be distributed in an office environment, and the robot was just switched on to do this task, it must know its actual position. Because the robot was just activated, it has no idea where it is. Therefore, it tries a self-localization that uses statistical methods like Monte Carlo. If the localization is successful, the robot has to drive to the post-office boxes. It knows the location by the use of the navigation map. As a rule the boxes are equipped with names. The robot must therefore be able to read the names, which assures a correct delivery. An OCR (optical character recognition) module must therefore be part of the navigation map. It shall be assumed that only one post-office box contains a letter. The robot then has to take the letter. The robot must also read the doorplates during the map acquisition, so it is able to determine to which address the letter must be brought. A path scheduler can then examine a beneficial run. If the robot has reached the desired office, it can place the letter on a desk, which should also be contained in the navigation map.

In Figure 1 is shown a possible architecture of a video-based robot-navigation program on a rather abstract level.

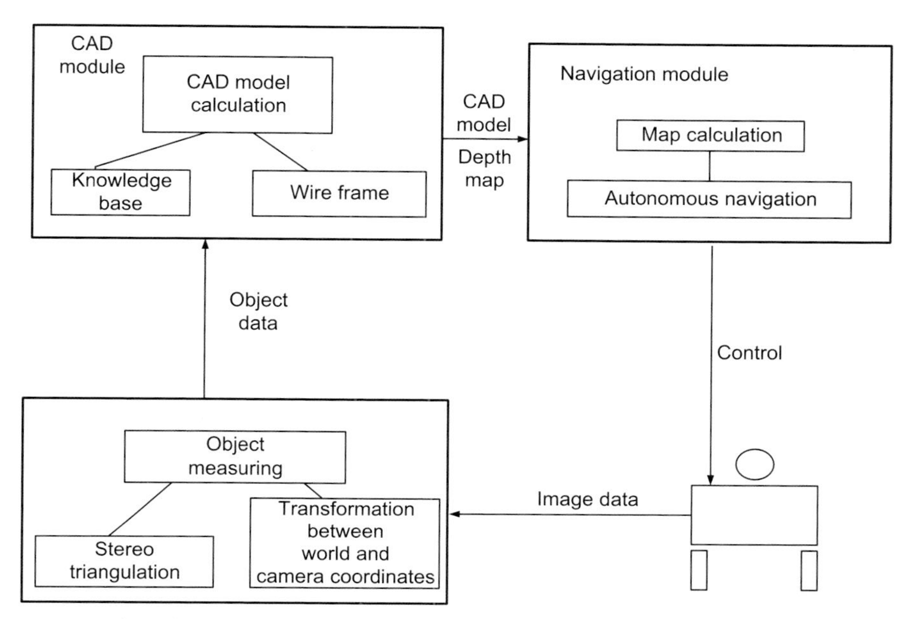


**Figure 1** The architecture of a video-based robot navigation software

A mobile robot, which is equipped with a camera, is sketched in the lower right area of the figure. The image data are read from an image-processing module that must try to detect an object in the image using image-processing operators. The object is analyzed and reconstructed after a successful detection. It is necessary to determine its world coordinates. The three-dimensional world coordinate system is independent of the robot's actual position. Its origin can be arbitrarily chosen. For example, the origin could be that point from which a robot starts its interior exploration. The origin of the three-dimensional camera coordinate system is determined by the focal point of the camera. If object coordinates are actually known in the camera coordinate system, it is possible to derive the world coordinates. The determination of the coordinates can use a stereo technique. At least two images from different positions are necessary for these purposes. Corresponding pixels belonging to that image region, which represents the desired object, must be detected in both images. Stereo triangulation exploits geometrical realities to determine the distance of the object point from the focal point. Additionally, the technical data of the camera must be considered for the depth estimation. Calibration techniques are available for these purposes. If the object coordinates are known, the object and its parts can be measured. In many cases the robot will be forced to take many more than two images for a reliable three-dimensional reconstruction, because three-dimensional objects often look different when viewed from different positions. The acquired data should then enable a CAD (computer-aided design) model to be produced. This can be a wire frame represented with a graph. For example, if the CAD model of an office table is to be obtained, the table legs and the desktop can be represented with edges. The program must determine for every edge the length and its start and end-points, which are represented by nodes. Coordinates are then attached to every node. The CAD module can additionally use a knowledge base for the proper reconstruction of the object. For example, the knowledge base can supply an image-processing program with important information about the configuration and quantity of object parts like the desktop and the table legs. After the three-dimensional object reconstruction is completed, the examined data can be collected in the navigation map. All these tasks must take place before the operating phase can start. The autonomous navigation uses the calculated map and transmits control commands to the mobile robot to fulfill the work necessary that depends on the particular scenario.

As noted before, such a service robot must be producible at a very low price if it is to fulfill its tasks cost effectively. Beside the use of very cheap equipment, the aim can be realized with the creation of favorable software. In particular, the software should be portable, work on different operating systems, and be easily maintainable.

Chapter two discusses some image-processing operators after these introductory words. The purpose of the chapter is not to give a complete overview about existing operators. Several textbooks are available. Image-processing operators are discussed that seem appropriate for machine-vision tasks. Most of the operators explained are used in experiments to test their presumed eligibility. Although cheap color cameras are available, the exploitation of color features in machine-vision applications is not often observed. Different color models are explained with regard to their possible field of application in machine-vision applications following elementary elucidations.

There then follows a section that relates to Kalman filter that is not a pure image-processing operator. In fact the Kalman filter is a stochastic method that can basically be used in many application areas. The Kalman filter supports the estimation of a model's state by the use of appropriate model parameters. Machine-vision applications can use the Kalman filter for the three-dimensional reconstruction by analyzing an image sequence that shows the object at different points in time. The image sequence can be acquired with a moving camera that effects the state transitions.

Video-based exploration with autonomous robots can be damaged by illumination fluctuations. The alterations can be effected by changes in the daylight, which can determine the robot's working conditions. For example, lighting alterations may be observable if the robot's working time comprises the entire day. Experiments showed that a Gabor filter can mitigate the effects of inhomogeneous illumination. The chapter discusses an application that uses the Gabor filter in a machine-vision application and reports the results.

Subsequent paragraphs describe fundamental morphological operators that are not typical for video-based machine-vision applications, but as a rule they are used in almost every image-processing program and thus also in experiments that are explained in the later chapters of this book. They are therefore explained for clarity. Further basis operators are edge detection, skeleton procedure, region building, and threshold operator. The skeleton procedure is not so frequently observed in machine-vision applications as the other listed operators, but it seems to be principally an appropriate technique if the three-dimensional reconstruction with wire frames is required. The skeleton procedure is therefore discussed for the sake of completeness.

Chapter three is devoted to navigation. Applications that control mobile service robots are often forced to use several coordinate systems. The camera's view can be realized with a three-dimensional coordinate system. Similar ideas can hold for a robot gripper when it belongs to the equipment of a mobile robot. Further coordinate systems are often necessary to represent the real world and the robot's view that is called the egocentric perception of the robot. Transformations between different coordinate systems are sometimes required. An example of this was mentioned before.

Map appearances can be coarsely divided into grid-based maps and graph-based maps. Graph-based maps are appropriate if quite an abstract modeling of the environment is to be realized. They offer the possibility that known algorithms for graphs can be used to execute a path plan between a starting point and an arrival point. For example, the path planning can be calculated on the condition that the shortest path should be found. Grid-based maps offer the possibility that the environment can be modeled as detailed as is wished. The grid technique was originally developed for maps used by human beings like city maps, atlases, and so forth.

After the discussion of several forms of grid-based maps, path planning is explained. The path length, the actual necessary behavior, and the abstraction level of the planning influence the path planning. One application is then exemplified that combines two abstraction levels of path planning.

The next section shows an example of an architecture that involves different map types. The chapter finishes with an explanation of the robot's self-localization.

Chapter four deals with vision systems. Machine vision is doubtless oriented to the human visual apparatus that is first illustrated. The similarity between the human visual apparatus and the technical vision system is then elaborated. To this belongs also behavior-based considerations like the attention control that determines how the next view is selected. Further sections consider interactions between observer and environment.

The remainder of chapter four explains current technical vision systems, which can be low priced. CMOS cameras are more expensive cameras. They are not considered because affordable development of mobile service robots is not possible with such cameras.

The content of chapter five is the three-dimensional reconstruction of objects. CAD techniques are especially considered, but other methods are also described. The application area for CAD techniques was originally industrial product development. The strategy for object reconstruction from image data differs therefore from the original application that uses CAD to model a new product, but object reconstruction uses image data to gain a CAD model from an existing object. Nevertheless, CAD techniques are appropriate for machine-vision applications. This is shown in the chapter. First, widespread CAD techniques are regarded and then followed by approximate modeling methods. Some models are a composite of different approaches. These are the hybrid models. Automated conversions between different models are proposed. One approach is then discussed that creates a CAD model from image data. The drawback of this is an elaborate calculation procedure. This is often observed if CAD models in machine-vision applications are used. But alternative approaches, whose purpose is not the explicit generation of a CAD model and sometimes not even a complete object reconstruction, also frequently suffer from this problem.

Knowledge-based approaches seem to be appropriate to diminish the calculation effort. The last application proposes a direct manipulation of the object, which is to be reconstructed, with marks. This strategy offers possibilities for simplification, but in some applications such marks may be felt to be disturbing. This may hold especially for applications with service robots, because human beings also use the robot's working environment. Mark-based procedures also require additional work or are impracticable. An application for a service robot scenario can not probably use the strategy, because too many objects have to be furnished with such marks.

Chapter six covers stereo vision that tries to gain depth information of the environment. The configuration of the used cameras provides geometrical facts, which can be used for the depth estimation. The task is the three-dimensional reconstruction of a scene point if only corresponding points in two or more images are known. The examination of corresponding points is sometimes very difficult, but this depends on the particular application. Three-dimensional reconstruction can also be gained from image sequences that were taken from a moving camera. In this case the Kalman filter can be used.

Chapter seven discusses the camera calibration that is a prerequisite for a successful reconstruction, because the camera parameters are determined with this strategy. The simplest calibration strategy is the pinhole camera calibration that determines only the camera's basis parameters like the focal length. But approaches also exist that consider further parameters. Lens distortion is an example of such parameters. Special calibration approaches exist for robot-vision systems. In this case the robot can be used to perform a self-calibration.

Several computer-vision applications use self-learning algorithms (Chapter 8), which can be realized with neural networks. OCR (Chapter 9) in computer vision is an example. Self-learning algorithms are useful here, because the appearance of the characters varies depending on the environment conditions. But changing fonts can also be a problem.

Until now the work can be considered as tutorial and shows that known methods are insufficient to develop a reliable video-based application for a mobile and autonomous service robot. In the next chapters methods are explained that will close the gap.

Chapter 10 proposes the use of redundant programs in robot-vision applications. Although redundant programming is, in general, a well-known technique and was originally developed to enhance the robustness of operating systems [2], it is not common to consider the use in computer-vision applications. First, the chapter describes the basics and elaborates general design guidelines for computer-vision applications that use the redundancy technique. The capability was tested with a robot-vision program that reads numbers on a doorplate. A high recognition rate was obtained.

A further drawback for a potential developer is the fact that no evaluation attempts can be found in the literature to compare different algorithms for service-robot applications. Chapter 11 reports on executed comparisons among algorithms. The algorithms are explained and then compared to experiment.

Chapter 12 explains a cost-effective calibration program that is based on pinhole-camera calibration. Most existing implementations use commercial software packages. This restricts the portability and increases the costs for licenses. In contrast the proposed implementation does not show these drawbacks. Additionally it is shown how a precise calibration object can be simply and cheaply developed.

Chapter 13 shows the superiority of the redundant programming technique in the three-dimensional reconstruction by the use of the CAD technique. A new CAD modeling method was developed for robot-vision applications that enables the distance-independent recognition of objects, but known drawbacks like mathematically elaborate procedures can not be observed. The CAD model extraction from image data with the new method is tested with a program. The results gained are reported. The sample images used were of extremely poor quality and taken with an off-the-shelf video camera with a low resolution. Nevertheless, the recognition results were impressive. Even a sophisticated but conventional computer-vision program will not readily achieve the reported recognition rate.

# 2
# Image Processing

Typically, an image-processing application consists of five steps. First, an image must be acquired. A digitized representation of the image is necessary for further processing. This is denoted with a two-dimensional function $\mathrm{I}(x, y)$ that is described with an array. $x$ marks a column and $y$ a row of the array. The domain for $x$ and $y$ depends on the maximal resolution of the image. If the image has size $n \times m$, whereby $n$ represents the number of rows and $m$ the number of columns, then it holds for $x$ that $0 \le x < m$, and for the $y$ analog, $0 \le y < n$. $x$ and $y$ are positive integers or zero. This holds also for the domain of I. $\mathrm{I}(x, y)_{\max}$ is the maximal value for the function value. This then provides the domain, $0 \le \mathrm{I}(x, y) \le \mathrm{I}(x, y)_{\max}$. Every possible discrete function value represents a gray value and is called a pixel. Subsequent preprocessing tries to eliminate disturbing effects. Examples are inhomogeneous illumination, noise, and movement detection.

If image-preprocessing algorithms like the movement detection are applied to an image, it is possible that image pixels of different objects with different properties are merged into regions, because they fulfill the criteria of the preprocessing algorithm. Therefore, a region can be considered as the accumulation of coherent pixels that must not have any similarities. These image regions or the whole image can be decomposed into segments. All contained pixels must be similar in these segments. Pixels will be assigned to objects in the segmentation phase, which is the third step [3]. If objects are isolated from the remainder of the image in the segmentation phase, feature values of these objects must be acquired in the fourth step. The features determined are used in the fifth and last step to perform the classification. This means that the detected objects are allocated to an object class if their measured feature values match to the object description. Examples for features are the object height, object width, compactness, and circularity.

A circular region has the compactness of one. The alteration of the region's length effects the alteration of the compactness value. The compactness becomes larger if the region's length rises. An empty region has value zero for the compactness. A circular region has the value one for circularity too. In contrast to the compactness, the value of the circularity falls if the region's length becomes smaller [4].

Image-processing libraries generally support steps one to four with operators. The classification can only be aided with frameworks.

*Robot Vision: Video-based Indoor Exploration with Autonomous and Mobile Robots.* Stefan Florczyk

ISBN: 3-527-40544-5

## 2.1
## Color Models

The process of vision by a human being is also controlled by colors. This happens subconsciously with signal colors. But a human being searches in some situations directly for specified colors to solve a problem [3]. The color attribute of an object can also be used in computer vision. This knowledge can help to solve a task [5, 6]. For example, a computer-vision application that is developed to detect people can use knowledge about the color of the skin for the detection. This can affect ambiguity in some situations. For example, an image that is taken from a human being who walks beside a carton, is difficult to detect, if the carton has a similar color to the color of the skin.

But there are more problems. The color attributes of objects can be affected by other objects due to light reflections of these objects [7]. Also colors of different objects that belong to the same class, can vary. For example, a European has a different skin color from an African although both belong to the class 'human being'. Color attributes like hue, saturation, intensity, and spectrum can be used to identify objects by its color [6, 8]. Alterations of these parameters can effect different reproductions of the same object. This is often very difficult to handle in computer-vision applications. Such alterations are as a rule for a human being no or only a small problem for recognition. The selection of an appropriate color space can help in computer vision. Several color spaces exist. Two often-used color spaces are now depicted. These are $RGB$ and $YUV$ color spaces. The $RGB$ color space consists of three color channels. These are the red, green, and blue channels. Every color is represented by its red, green, and blue parts. This coding follows the three-color theory of Gauss. A pixel's color part of a channel is often measured within the interval $[0; 255]$. Therefore, a color image consists of three gray images. The $RGB$ color space is not very stable with regard to alterations in the illumination, because the representation of a color with the $RGB$ color space contains no separation between the illumination and the color parts. If a computer-vision application, which performs image analysis on color images, is to be robust against alterations in illumination, the $YUV$ color space could be a better choice, because the color parts and the illumination are represented separately. The color representation happens only with two channels, $U$ and $V$. $Y$ channel measures the brightness. The conversion between the $RGB$ and the $YUV$ color space happens with a linear transformation [3]:

$$\begin{pmatrix} Y \\ U \\ V \end{pmatrix} = \begin{pmatrix} 0.299 & 0.587 & 0.114 \\ -0.147 & -0.289 & 0.436 \\ 0.615 & -0.514 & -0.101 \end{pmatrix} \begin{pmatrix} R \\ G \\ B \end{pmatrix}. \tag{2.1}$$

This yields the following equations [9]:

$$Y = 0.299R + 0.587G + 0.114B, \tag{2.2}$$

$$U = -0.147R - 0.289G + 0.436B, \tag{2.3}$$

$$V = 0.615R - 0.514G - 0.101B. \tag{2.4}$$

To show the robustness of the $YUV$ color space with regard to the illumination, the constant $c$ will be added to the $RGB$ color parts. Positive $c$ effects a brighter color impression and negative $c$ a darker color impression. The constant $c$ affects only the brightness $Y$ and not the color parts $U$ and $V$ in the $YUV$ color space if a transformation into the $YUV$ color space is performed [3]:

$$Y(R+c, G+c, B+c) = Y(R, G, B) + c, \tag{2.5}$$

$$U(R+c, G+c, B+c) = U(R, G, B), \tag{2.6}$$

$$V(R+c, G+c, B+c) = V(R, G, B). \tag{2.7}$$

The sum of the weights in Equations (2.3) and (2.4) is zero. Therefore, the value of the constant $c$ in the color parts is mutually cancelled. The addition of the constant $c$ is only represented in Equation (2.2). This shows that the alteration of the brightness effects an incorrect change in the color parts of the $RGB$ color space, whereas only the $Y$ part is affected in the $YUV$ color space. Examinations of different color spaces have shown that the robustness can be further improved if the color parts are normalized and the weights are varied. One of these color spaces, where this was applied, is the $(YUV)'$ color space, which is very similar to the $YUV$ color space. The transformation from the $RGB$ color space into the $(YUV)'$ color space is [3]:

$$\begin{pmatrix} Y' \\ U' \\ V' \end{pmatrix} = \begin{pmatrix} \frac{1}{3} & \frac{1}{3} & \frac{1}{3} \\ \frac{1}{2} & 0 & -\frac{1}{2} \\ -\frac{1}{2\sqrt{3}} & \frac{1}{\sqrt{3}} & -\frac{1}{2\sqrt{3}} \end{pmatrix} \begin{pmatrix} R \\ G \\ B \end{pmatrix}. \tag{2.8}$$

The explanations show that the $YUV$ color space should be preferred for object detection by the use of the color attribute if the computer-vision application has to deal with changes in the illumination [3].

## 2.2 Filtering

### 2.2.1 Kalman Filter

A Kalman filter can be used for state estimation in dynamic systems. It is a stochastic filter [10]. The following description of the discrete linear Kalman filter is based on [11, 12]. The estimation is frequently based on a sequence of measurements, which are often imprecise. A state estimation will be found for which a defined estimate error is minimized. It is possible to estimate a state vector on the basis of measurements in the past with a Kalman filter. It is also possible to predict a state vector in the future with a Kalman filter. State vector $q(t+1)$, whereby $t$ denotes a point in the time-series measurements with $t = 0, 1, \ldots$, can be processed by the addition of random vector $\omega(t)$ to the product of the state vector $q(t)$ with the transition matrix $\Phi(t+1;t)$. $\Phi(t_2;t_1)$ denotes the transition from time $t_1$ to time $t_2$. $\Phi(t;t)$ is the unit matrix:

$$q(t+1) = \Phi(t+1;t)q(t) + \omega(t). \tag{2.9}$$

The observation vector $o(t)$ can be processed with the state vector $q(t)$, the observation matrix $H(t)$, and also the random vector $r(t)$:

$$o(t) = H(t)q(t) + r(t). \tag{2.10}$$

The two random vectors $\omega(t)$ and $r(t)$ have an expectation value of 0, $\mathrm{E}[\omega(t)] = \mathrm{E}[r(t)] = 0$. Both vectors are uncorrelated and have known statistical attributes. The following formula is valid for a linear state valuer:

$$\tilde{q}(t+1) = \Phi(t+1;t)^{*}\tilde{q}(t) + K(t)o(t). \tag{2.11}$$

$\Phi(t+1;t)^{*}$ and $K(t)$ are matrices that must be determined under an unbiased valuer:

$$\mathrm{E}[q(t)] = \mathrm{E}[\tilde{q}(t)]. \tag{2.12}$$

After some insertions and transformations, the following equation for $\Phi(t+1;t)^{*}$ holds:

$$\Phi(t+1;t)^{*} = \Phi(t+1;t) - K(t)H(t). \tag{2.13}$$

Further insertions yield an equation for $\tilde{q}(t+1)$ at time $t+1$:

$$\tilde{q}(t+1) = \Phi(t+1;t)\tilde{q}(t) + K(t)(o(t) - H(t)\tilde{q}(t)). \tag{2.14}$$

The matrix $K(t)$ can be calculated by fulfilling the demand that the expected value of estimation error $\varepsilon(t) = [q(t) - \tilde{q}(t)]$ must be minimal. Two estimations are performed. The error is measured for every estimation. The system state at time $t$ over the state transition equation $\tilde{q}(t+1)^{-}$ is one of these two estimations:

$$\tilde{q}(t+1)^{-} = \Phi(t+1;t)\,\tilde{q}(t). \tag{2.15}$$

The second estimation $\tilde{q}(t+1)^+$ denotes the improved estimation that takes place on the basis of the observation $o(t+1)$:

$$\tilde{q}(t+1)+ = \Phi(t+1;t)\tilde{q}(t)+ + K(t)[o(t+1) - H(t+1)\tilde{q}(t+1)^-]. \tag{2.16}$$

Estimation errors $\varepsilon(t)^-$ and $\varepsilon(t)^+$ can be processed with associated covariance matrices $P(t)^-$ and $P(t)^+$.

### 2.2.2 Gabor Filter

A Gabor filter belongs to the set of bandpass filters. The explanation begins with the one-dimensional Gabor filter and follows [13]. The spectrum of a bandpass filter specifies the frequencies that may pass the filter. The middle frequency is a further parameter. The impulse answer of the one-dimensional analytical Gabor filter is given in:

$$g(x) = \frac{1}{\sqrt{2\pi\sigma}} e^{-x^2/2\sigma^2} e^{if_{X_m} x}. \tag{2.17}$$

$f_{X_m}$ specifies the middle frequency and $\sigma^2$ the variance of the included Gauss function. i represents an imaginary number. If the Fourier transformation $F\{g(x)\}$ is applied to the Gabor filter $g(x)$ in the spatial domain, the following formula holds:

$$F\{g(x)\} = G(f_X) = \frac{1}{\sqrt{2\sigma}} (e^{-\frac{\sigma^2}{2}(f_X - f_{X_m})^2}). \tag{2.18}$$

$N = \frac{1}{\sqrt{2\pi\sigma}}$ is derived by the normalization of the included Gauss function:

$$N \int_{-\infty}^{\infty} e^{-x^2/2\sigma^2} dx = 1. \tag{2.19}$$

Other kinds of normalizations exist. But the choice of the normalization has no effect on the calculation of the Gabor filter's local phase $\varphi(x)$:

$$\varphi(x) = f_{X_m} x. \tag{2.20}$$

The local phase of a bandpass filter yields local information in terms of distance to an origin $O$ in a coordinate system. The real part of the Gabor filter is also known as the even Gabor filter and the imaginary part as the odd Gabor filter. The modulus of the impulse answer is now given:

$$|g(x)| = e^{-x^2/2\sigma^2}. \tag{2.21}$$

The even and the odd Gabor filters are shown in the left side of Figure 2. The right side depicts the amount and the local phase.

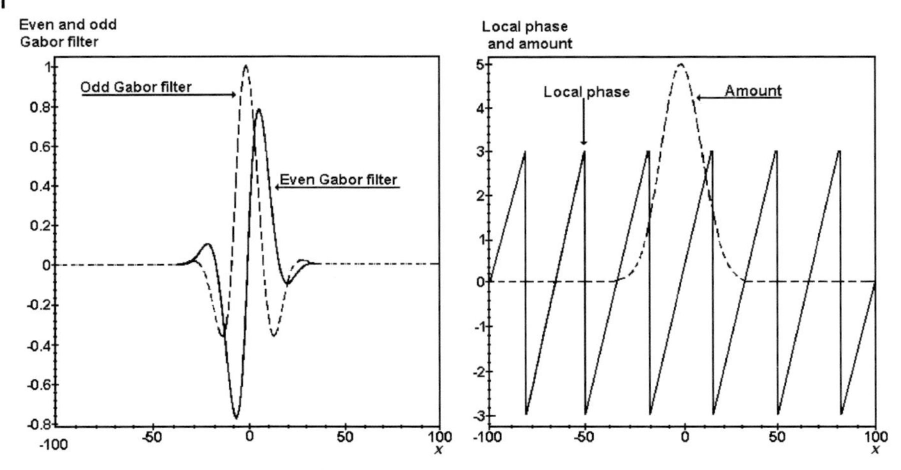


**Figure 2** The one-dimensional Gabor filter [13]

The local phase is not affected by the amount but by the ratio of the even and the odd Gabor filter. The Gabor filter in the figure can be generated with $f_{X_m} = 0.19$ and $\sigma = 10.5$. It can be seen with an appropriate confidence measurement if the local phase of the Gabor filter is stable on real images with noise. The stability is given if the confidence measurement is fulfilled from the amount of the impulse answer. The Gabor filter can also be written by using two parameters, the Gabor wavelength $\lambda$ and a spectrum factor $s$:

$$\lambda = \frac{2\pi}{f_{X_m}} \qquad s = \frac{1}{f_{X_m}\sigma} = \frac{\lambda}{2\pi\sigma}. \tag{2.22}$$

The Impulse answer of the Gabor filter is given with these two parameters:

$$g(x) = \frac{\sqrt{2\pi}s}{\lambda} \cdot e^{-2(\frac{\pi s}{\lambda})^2 x^2} e^{i\frac{2\pi}{\lambda}x}. \tag{2.23}$$

If the spectrum factor $s$ remains constant, and only the Gabor wavelength $\lambda$ changes, the form and spectrum of the resulting Gabor filters are equal. However, the alteration of the spectrum factor $s$ yields different Gabor filters, because the number of frequencies decreases with an increasing spectrum factor. This can be seen in Figure 3.

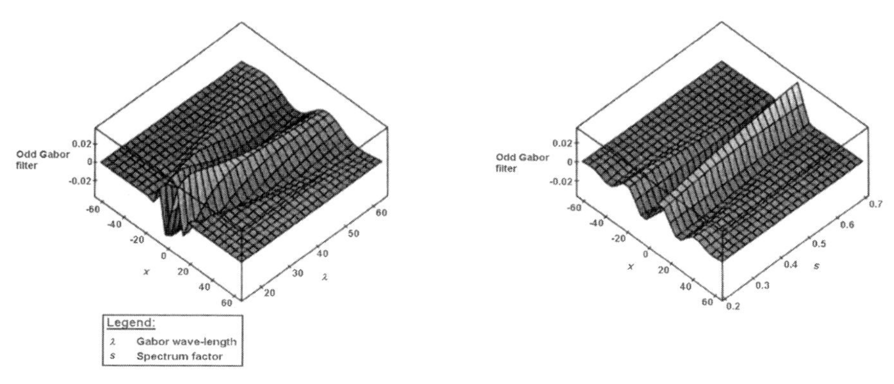


**Figure 3** The variation of Gabor wavelength and spectrum factor [13]

The figure shows the odd part of the Gabor filter. The variation of the Gabor wavelength $\lambda$ by a constant spectrum factor $s = 0.33$ is illustrated in the left part of the drawing. The variation is performed in the interval [14; 64]. It can be seen that no change happens to the form of the Gabor filter. The Gabor wavelength, $\lambda = 33$ pixels, is kept constant in the right side of Figure 3, and the spectrum factor $s$ varies in the range [0.2; 0.7]. It can be observed that the number of frequencies is changing with the alteration of the spectrum factor. The variation of the spectrum in the given interval $[\lambda_{\min}; \lambda_{\max}]$ in pixels is shown in the following. The spectrum of the filter can be measured in octaves $o$. The next equation shows the relation between $o$ and the spectrum factor $s$:

$$s = \frac{2^o - 1}{2^o + 1} \quad \Leftrightarrow \quad o = \log_2\left(\frac{1+s}{1-s}\right). \tag{2.24}$$

For example, if the spectrum factor is one octave ($s = 0.33$) with Gabor wavelength $\lambda = 33$ pixels, then the interval of the spectrum $[\lambda_{\min}; \lambda_{\max}]$ has the values [23; 53]. The spectrum interval can be calculated with

$$\lambda_{\min,\max} = \frac{\lambda}{1 \pm s\sqrt{-2\ln(a)}}. \tag{2.25}$$

$a$ is that part of the amplitude that is least transmitted within the filter spectrum. It holds that $a \in [0; 1]$ with $|\tilde{G}(\lambda_{\min,\max})| = a \cdot |\tilde{G}(\lambda)|$, whereas $\tilde{G}(\lambda)$ is equivalent to the replaced Fourier transformed function $G(f_X)$.

We now show the impulse answer of a two-dimensional Gabor filter as explained in [14]:

$$g(x, y) = e^{-\frac{x^2+\delta^2 y^2}{2\sigma^2}} \left(e^{-2\pi i \frac{x}{\lambda}} - e^{-\frac{\sigma^2}{2\delta}}\right). \tag{2.26}$$

The included Gauss function has, in the two-dimensional case, width $\sigma$ in the longitudinal direction and in the cross direction a width $\sigma/\delta$. $\delta$ denotes the dilation. The last constant term is necessary to obtain invariance for the displacement of the gray-value intensity of an image. The formula is also known as the mother wavelet. The complete family of similar daughter wavelets can be generated with the following equations:

$$g(x, y) = 2^{-2m} g(x', y'), \tag{2.27}$$

$$x' = 2^{-m} [x\cos(\theta) + y\sin(\theta)] - \tau, \tag{2.28}$$

$$y' = 2^{-m} [-x\sin(\theta) + y\cos(\theta)] - \tau'. \tag{2.29}$$

The parameter $m$ is an integer value and marks the extension and the frequency of the wave. The translation is labeled with $\tau$, $\tau'$, and the rotation with $\theta$.

### 2.2.3
### Application of the Gabor Filter

The Gabor filter was tested in a computer-vision project [14] in which a wooden cube had to be grasped with a robot gripper. Images were taken with a hand camera. The cube is part of a montage process for wooden parts that must be assembled to aggregates. The assembly procedure can be decomposed into several steps. First, the required component must be identified. Once this has been done it is necessary to move the robot gripper to the detected object. Then the fine localization of the gripper takes place. The object is then gripped and used for the montage process. It is highly important that the localization of the object must be very accurate, because an inexact localization can lead to problems in the following steps of the montage process. For example, it may be possible that the entire montage process fails. Because of the real-world aspect of this application, it is expected that it will not be confused by alterations in the illumination. Three parameters are necessary for the localization of the object. These can be the two-dimensional $(x, y)$ position of the object and vertical angle of rotation $\theta$. The controller realizes the object localization and is responsible for calculating correction movement $(\Delta x, \Delta y, \Delta \theta)$. This controller is implemented with self-learning algorithms. The required correction movement can not be processed generally in one step, several steps are required.

The recognition of the wooden cube, which served as the test object, is not as simple as it seems at first sight. The wooden cube has strongly rounded edges. This can result in an oval shape if the images are taken from a rather inclined view. Three axial thread axes are the second problem, which affect a strong shadow inside the cube. So it can be difficult to detect the wooden cube that is included in a set of other objects.

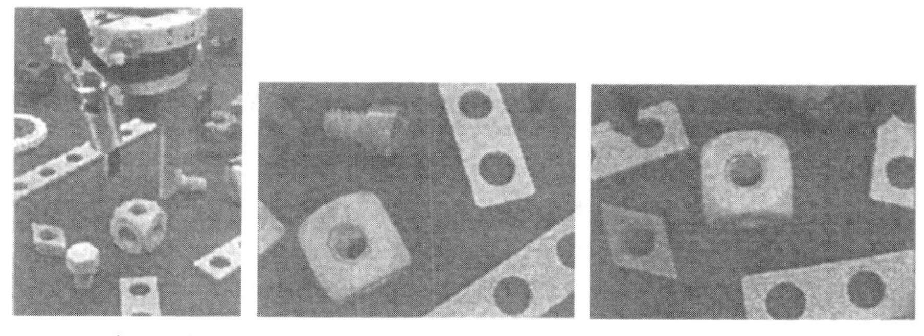

**Figure 4** The wooden cube within a set of other objects [14]

The left part of Figure 4 shows the gripper above several wooden parts. In the middle part of the figure the gripper was moved above the wooden cube. The right part shows an image that has been taken after the fine positioning of the gripper. The two problems can result in a wrong detection of another object. The tests were performed with images taken from a hand camera. The recognition of the wooden cube starts with taking the camera image, which is then preprocessed, see Figure 5.

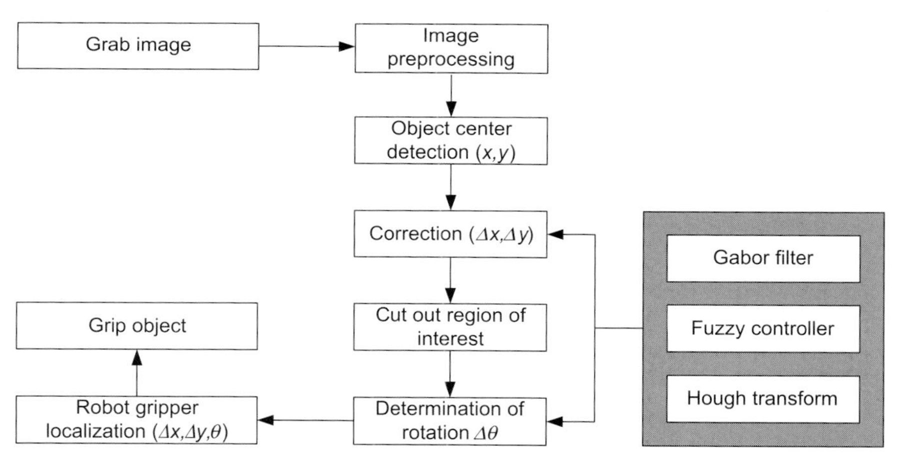


**Figure 5** The regulator circle [14]

After preprocessing, the object center $(x, y)$ is calculated. The calculated object center is then tested with a Gabor filter or further approaches like the Hough transformation [15] or a fuzzy controller. The test provides the necessary correction $(\Delta x, \Delta y)$ of the object center. Then the region of interest (ROI) is cut from the entire image that contains the wooden cube in the center of the image clip. In the next step a further test is applied that also uses the Gabor filter or the two other approaches. These are the Hough transformation or the fuzzy controller as mentioned before. In this step the required angle of rotation $\theta$ is determined. Now, the necessary information exists to do the localization of the robot gripper, which receives the movement values $(\Delta x, \Delta y, \Delta\theta)$.

The Gabor filter and the Hough transformation are part of a self-learning algorithm. The taking of the test images and training images and the calculation of the parameters is realized with an automated process. For these purposes the training of the controller happens with a single demonstration of the optimal grip position. The use of the Gabor filter in the application is shown now in more detail. The calculation of the wooden cube center $(x, y)$ is, in comparison to the examination of the angle of rotation $\theta$, rather simple, because it is often difficult to recognize the cube form. To handle this problem, the Gabor filter is used for the representation of the wooden cube. The stimulation for the representation of the wooden cube with the Gabor filter was the work of Jones and Palmer. They showed by experiments with cats that the impulse answer of so-called simple cells in the cat's visual cortex can be approximated with the model based on Gabor filters as shown before [16]. The approximation is shown in Figure 6.

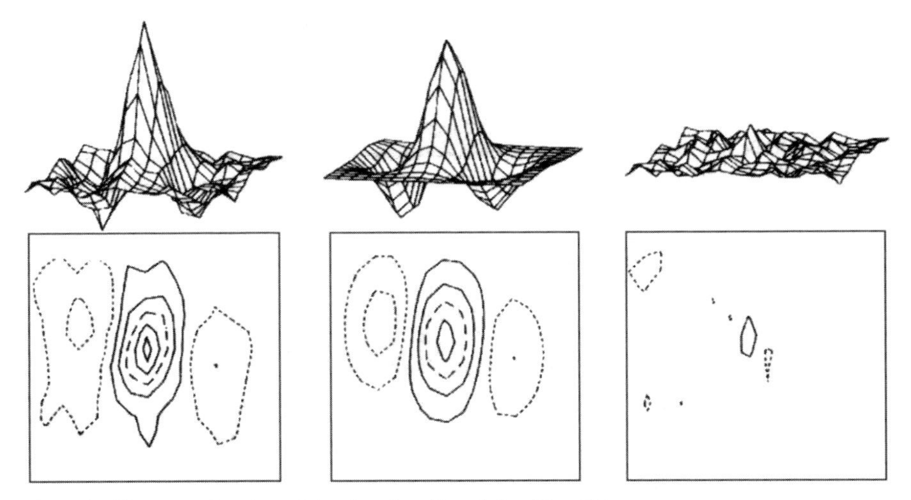

**Figure 6** The approximation of simple cells with a Gabor filter [16]

The left side of Figure 6 shows the impulse answer from the simple cells that was determined with experiments. This impulse answer is adapted with a Gabor filter, which is shown in the middle of Figure 6. The difference between the experimentally determined impulse answer and the approximated Gabor filter is depicted in the right side of Figure 6 [16].

A simple cell can now be modeled with a two-dimensional Gabor function that has the center $(\tau, \tau')$, the wavelength $\lambda/2^m$, the direction $\theta$, and an included Gauss function with the extensions $\sigma$ and $\sigma/\delta$. The system uses $n$ neurons that are all looking at the same image and represent a model of a simple cell. The difference between these neurons can be found in their centers $(\tau, \tau')$. The parameters of the

Gabor function were not known at the beginning of the system design. Therefore, tests were necessary that have been performed with $k$ different Gabor functions. The measurement of the test error was determined by the use of the precision in the prediction of the object orientation, which is noted with the angle $\theta$. Values for the three parameters $\delta$, $\sigma$ and $\lambda$ had to be found. Therefore, nine coordinates $(\tau, \tau')$ that were continuously distributed on that image were chosen from an image.

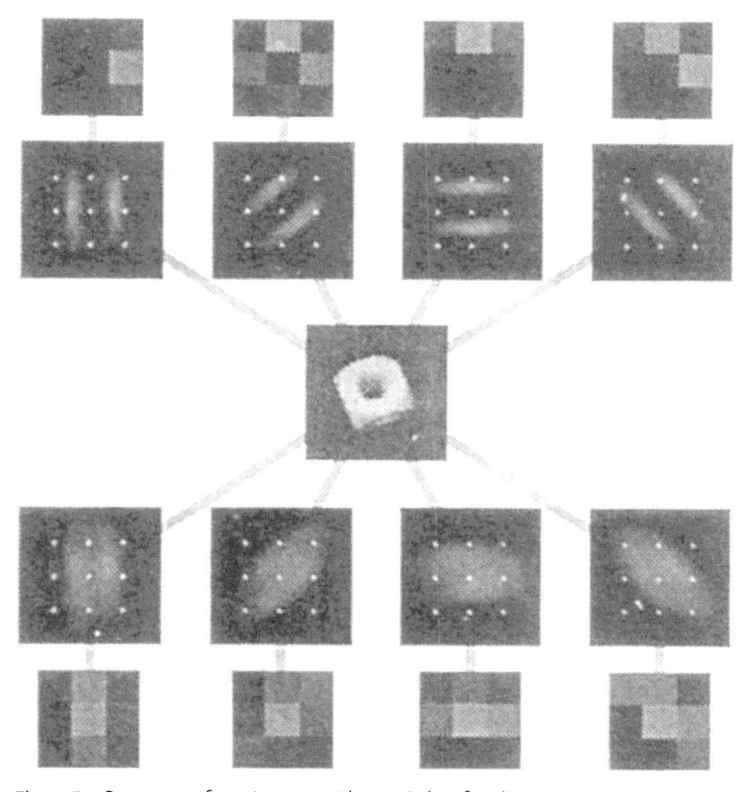

**Figure 7** Sequence of test images with two Gabor families [14]

Figure 7 shows the wooden cube. Two different Gabor families have been applied to the cube. Four orientations $\theta$ and the nine chosen coordinates can be seen in the images. The parameters $\delta = 1.25$, $\sigma = 1.25$, and $\lambda = 0.15$ were used in the top part of Figure 7, whereas the values $\delta = 1.00$, $\sigma = 1.50$, and $\lambda = 0.10$ are valid in the lower part of Figure 7. The scanning of $\delta$, $\sigma$, and $\lambda$ was performed for four orientations. This results in 36 $(9 \times 4)$ values for $\delta$, $\sigma$, and $\lambda$. A neural network [17] received these 36 values per image as a training set. The output of the neural network is the calculated rotation angle $\theta$. The selection of the parameters $\delta$, $\sigma$, and $\lambda$ happens after $k$ training runs by the minimization of the test error in the output of the angle $\theta$.

The wooden cube has a fourfold symmetry. Images with the orientation $\theta = 45^\circ$ and $\theta = -45^\circ$ are identical. The neural network should be able to recognize this fact. This problem can be solved with a $c$-fold pair coding with $\sin(c\theta)$ and $\cos(c\theta)$. Additionally two neurons in the output layer were used instead of one neuron. The rotation angle $\theta$ can be calculated with $\theta = \arctan(\sin(c\theta)/\cos(c\theta))/s$. The result's quality can also be controlled with the number of neurons in the hidden layer. Good results were gained with combination that had 36 neurons in the input layer, nine neurons in the hidden layer, and two neurons in the output layer.

The approach, which is based on the Gabor filter, should also be able to detect wooden cubes of different colors. This problem can be solved by the approximation of brightness and contrast to the original training conditions. So it can be possible to handle the problem of color alterations as well as alterations in illumination. The hand camera yields an $RGB$ color image. This will be converted into a gray image by maximization of the color channels for every pixel $p^i = \max(R^i, G^i, B^i)$. Deviations in contrast and illumination have been also adapted to the conditions that were valid during the training phase:

$$p^{i} = (p^{i} + c_1)\cdot c_2 \quad \text{with} \quad c_1 = \iota_m^{I}\cdot\sigma^{I'}/\sigma^{I} - \iota_m^{I'} \quad \text{and} \quad c_2 = \sigma^{I}/\sigma^{I'},$$

$$\text{whereby} \quad \iota_m^{I} = \text{mean}\,(\text{I}) \quad \text{and} \quad \sigma^{I} = \text{std}\,(\text{I}). \tag{2.30}$$

$\iota_m^{I}$ is the middle intensity and $\sigma^{I}$ the standard deviation of a gray image I during the training phase. I′ is an actual image that has to be analyzed. It was possible to grasp wooden cubes by candlelight. The grasping process was robust without the necessity for additional training effort. Figure 8 shows some examples. The wooden cube could be detected and grasped.

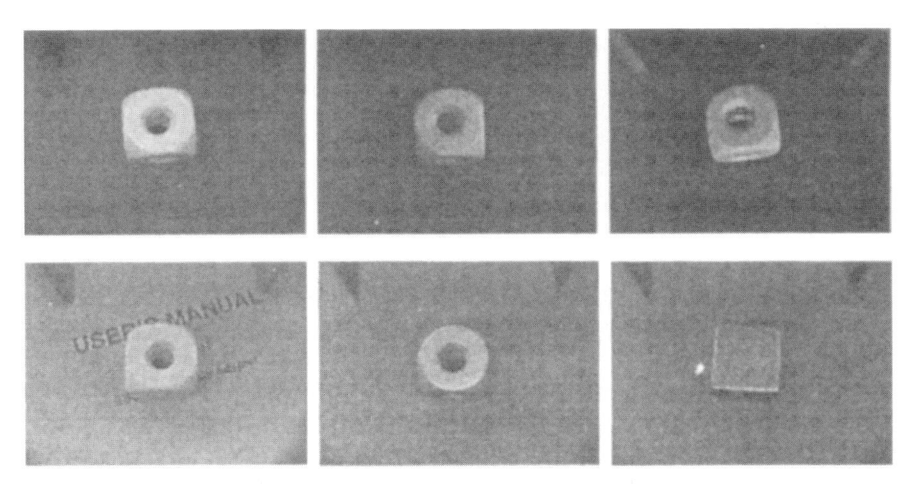


**Figure 8** The wooden cube under different conditions [14]

The top-left image shows illumination conditions during the training phase, the top-middle image weak illumination, the top-right image glaring light, the lower-left image shows a textured working area, the lower-middle image a blue cube, and the lower-right image shows a red square stone.

The control cycle of the Gabor system starts with the calculation of the central point of a normalized hand-camera image. The necessary correction movement $\Delta x$ and $\Delta y$ is calculated with the first neural network. It is problem if a wooden cube is positioned at the border of an image. This must be recognized because of the need for a new image. Then an image clip, which contains a wooden cube, is cut out. The Gabor filter is then applied to this image clip. The calculated result, which is represented by the orientation angle $\theta$, is provided for the second neural network. Now the correction movement of the robot gripper is performed, before the process of gripping is executed.

Experiments with the system have shown that the precise positioning of the robot gripper depends on the training of the neural networks. The wooden cube was put nearby the optimal position of the robot gripper (start position). Then a randomly chosen translation and rotation of the robot gripper was effected. A sequence of $n$ steps for the fine positioning was applied. An image is taken, the required translation and rotation calculated, then applied, and the position and orientation of the robot gripper recorded in every step. The robot gripper returns to the start position after $n$ steps and starts again with further randomly executed translation and rotation. These $n$ steps were repeated $N$ times. Now it is possible to determine for each of the $n$ steps the standard deviation relating to the start position. These $N$ tries were repeated several times to exclude systematical effects, which can result from the choice of the start position. Another start position is chosen in every new try.

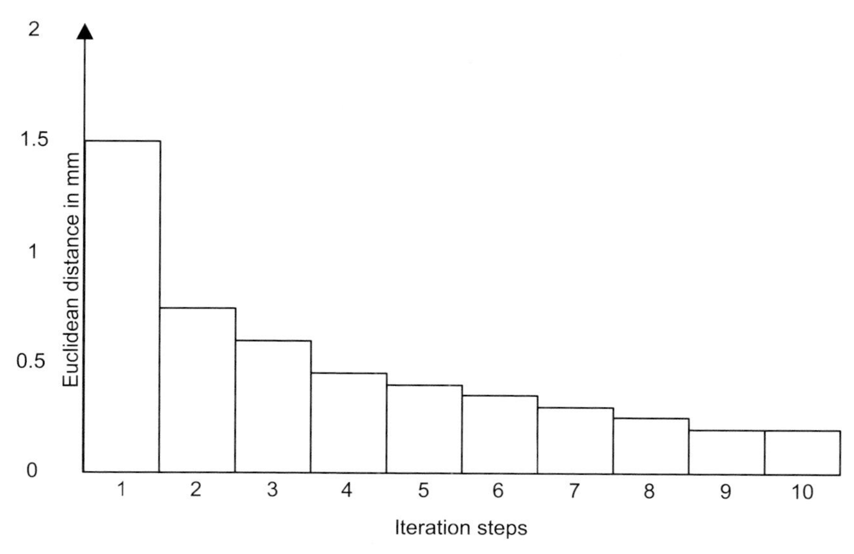


**Figure 9** Gripping precision with the Gabor approach [14]

The increase of the precision in the gripping position with the number of steps can be seen in Figure 9. The figure shows the middle Euclidean error on the $Y$-axis. The number of steps can be read from the $X$-axis. The middle Euclidean error has the value 1.5 mm in the first step. The middle Euclidean error remains under 0.5 mm as of the 4th step.

## 2.3 Morphological Image Processing

Morphological image processing is based on mathematical set theory and can be applied to gray and binary images. The demonstration of the morphological image processing in this work is mainly based on [18]. Only the morphological image processing on binary images is discussed here. The extraction of image regions is supported with a structuring element. A structuring element can have different forms. These can be, for instance, a circle, a rectangle, or a rhombus. The selection of the form depends on the objects in the image to which the structuring element is to be applied and the contents of the image, for example, noise that should be suppressed.

Erosion and dilation are the basis of the morphological image processing. All operators that are known in morphological image processing are constructed from these two base operators.

### 2.3.1 The Structuring Element

The structuring element is a small set that is applied to an image. Every structuring element has a reference point. If a structuring element is placed on a binary image, it is checked whether the pixel that is covered from the reference point is set. If this is true, the structuring element can be applied. Figure 10 shows some structuring elements.

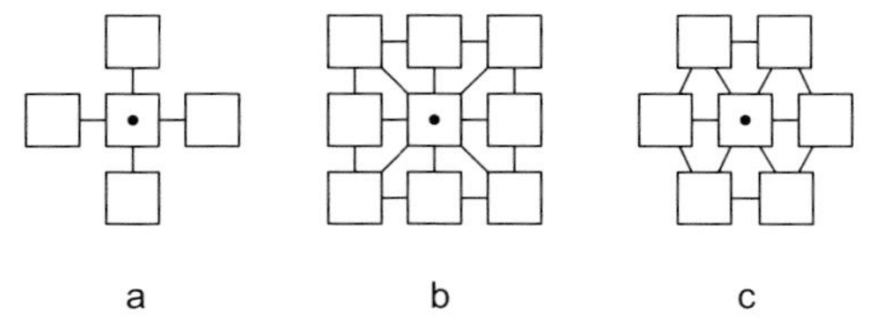


**Figure 10** Some structuring elements [18]

The structuring elements are constructed from the squares that represent pixels. The square that has a dot in the middle pixel is the reference point. Figure 10 shows a rhombus (a), a square (b), and a hexagon (c) as structuring elements, which are constituted from the particular pixels. The form and the size of the structuring element depend on the objects in the image to which the structuring element is to be applied.

### 2.3.2 Erosion

Erosion tests if the structuring element fits completely into a pixel set. If this is true, the pixel set constructs the eroded set. All pixels that belong to the considered pixel set are transformed to the reference point. Only binary images are considered here. This can be formulated mathematically. The erosion of a pixel set $A$ with a structuring element $S$ is denoted with $\mathrm{E}_S(A)$. This is the set of $n$ pixels $p^i$, $i = 1, 2, \ldots, n$, for which $S$ is completely contained in $A$ if the reference point is positioned at the point $p^i$:

$$\mathrm{E}_S(A) = \{p^i \,|\, S \subseteq A \; 1 \leq i \leq n\}. \tag{2.31}$$

Figure 11 shows an example of the erosion.

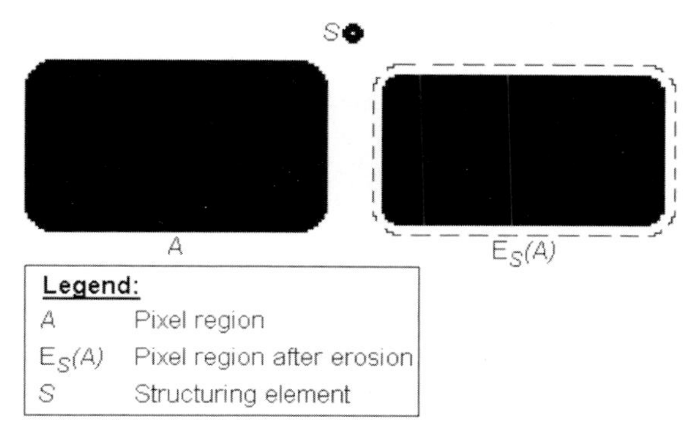


**Figure 11** The erosion of the set $A$

Pixel region $A$ is pictured in the left part of Figure 11. Erosion was performed by applying the circular structuring element $S$ to $A$ that yields the result $\mathrm{E}_S(A)$ that is represented by the black area in the right part of the figure. The zone between the dashed line and $\mathrm{E}_S(A)$ is eliminated.

### 2.3.3 Dilation

Dilation produces an opposite effect in comparison to erosion. It extends coherent pixel sets. The magnitude of the enlargement is controlled by the size of the structuring element used. The larger the used structuring element the larger is the effected extension. A useful effect of the dilation is the merging of regions if they are close together and if the size of the structuring element has been determined accordingly. Before the dilation is demonstrated with an example, a more formal description is introduced:

$$\delta_S(A) = \{p^i \mid S \cap A \neq 0 \; 1 \leq i \leq n\}. \tag{2.32}$$

The application of the dilation to a set $A$, which consists of $n$ pixels $p^i$, with a structuring element $S$ provides a new set of pixels $\delta_S(A)$.

The application of the dilation means that every pixel in the original set $A$ is transformed to the shape of the structuring element on condition that the actually considered pixel of the original set represents the reference point of the structuring element $S$ [19].

Figure 12 shows an example of the dilation.

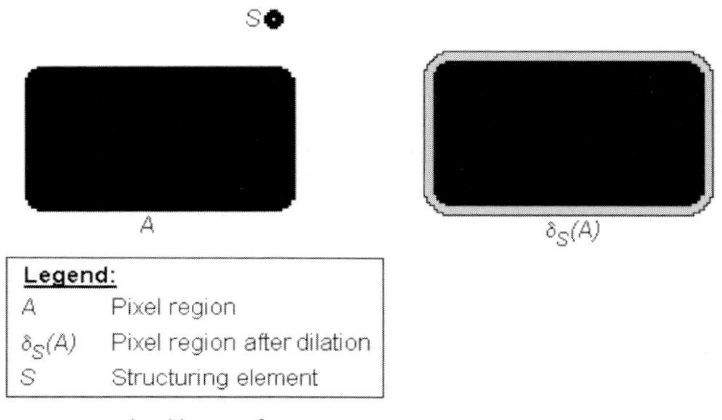


**Figure 12** The dilation of a set $A$

The original set $A$ is transformed with the structuring element. The effect of the dilation is visible at the border of the original set. The result shown can be gained by moving the structuring element along the border of the original set on condition that the coordinates of the actually considered pixel in the original set and the reference pixel of the structuring element are the same [19].

## 2.4 Edge Detection

To detect edges, it is not sufficient to analyze only the pixels. Rather it is necessary to include the entire neighborhood in the inspection. This strategy will find gray-value jumps, which can be observed if edges exist. Searching for the gray-value jumps will help edge detection. Edge operators like the Sobel operator use convolution to obtain the edges of an image. The mathematical derivation and explanation of the Sobel filter is now shown according to [19].

An original image I is transformed with the convolution into image I′ with the help of matrix $K$. The matrix $K$ is moved over the original image I, and a multiplication is executed for every value in the matrix $K$ with the respectively covered value in the matrix I. All results of the performed multiplications are added to a new value, which is written into the cell of the new matrix I′ that has the same position as the cell in the original image I covered from the middle cell of the matrix $K$.

| | | | | | | | | | | |
|---|---|---|---|---|---|---|---|---|---|---|
| | | | 3 | 4 | 7 | 12 | | | | |
| 77 | 89 | | 1 | 3 | 2 | 18 | | 1 | 1 | 1 |
| 111 | 125 | | 19 | 5 | 6 | 14 | | 1 | 10 | 1 |
| I′ | | | 20 | 6 | 4 | 13 | | 1 | 1 | 1 |
| | | | | I | | | | | $K$ | |

**Figure 13** Example for a convolution

The convolution procedure is demonstrated in Figure 13. The filter mask used $K$ is of size $3 \times 3$. Every cell in the matrix has the value one except for the center that has the value 10. This mask was moved over the image I to examine the four values in the matrix I′. For example, the value 77 was calculated as the matrix $K$ was covering the nine values in the first three rows and first three columns in the matrix I:

$$77 = 3 + 4 + 7 + 1 + 3 \times 10 + 2 + 19 + 5 + 6. \tag{2.33}$$

The explained strategy can not be used to determine new gray values for numbers that can be found at the border of the image I. The original values can be adopted unchanged into the new image I′ or they can be omitted in the new image I′ as is done in Figure 13. A domain restricts gray values. Often the domain is determined with the interval $[0; 255]$. The particular valid domain for the gray values can be exceeded in the new matrix I′ if the convolution is executed. Therefore, it is necessary to multiply the new calculated values with a constant $c_0$. To avoid negative values in the new matrix I′, it is also necessary to add a further constant value $c_1$:

$$\mathrm{I}'(x,y) = c_0 \sum_{i=-k}^{+k} \sum_{j=-l}^{+l} K(i,j)\mathrm{I}(x-i, y-i) + c_1. \tag{2.34}$$

Coefficients $K(i,j)$ for the convolution can be found in the matrix $K$ with $2k + 1$ columns and $2l + 1$ rows.

An edge can be detected in gray-value images by searching for gray-value jumps. So the neighborhood of pixels must be examined. To get an understanding of the functionality of edge operators like Sobel operator, the first derivative of an image function is shown here:

$$\mathrm{I}'_x(x,y) = \frac{\partial \mathrm{I}(x,y)}{\partial x} \quad \text{and} \quad \mathrm{I}'_y(x,y) = \frac{\partial \mathrm{I}(x,y)}{\partial y}. \tag{2.35}$$

The image function has two dimensions. So it is necessary to compute derivatives for the $x$ and $y$ variables. The discrete variants of the continuous derivatives will be processed, because an image is a two-dimensional discrete array:

$$\frac{\partial \mathrm{I}(x,y)}{\partial x} \approx \Delta x \mathrm{I}(x,y) = \frac{\mathrm{I}(x,y) - \mathrm{I}(x - \Delta x, y)}{\Delta x} \tag{2.36}$$

and

$$\frac{\partial \mathrm{I}(x,y)}{\partial y} \approx \Delta y \mathrm{I}(x,y) = \frac{\mathrm{I}(x,y) - \mathrm{I}(x, y - \Delta y)}{\Delta y}. \tag{2.37}$$

These two formulas are the discrete derivatives for the $x$ and $y$ variables. When $\Delta_x = 1$ and $\Delta_y = 1$ the derivatives for $x$ and $y$ are written as:

$$\Delta_x \mathrm{I}(x,y) = \mathrm{I}(x,y) - \mathrm{I}(x-1,y), \tag{2.38}$$

$$\Delta_y \mathrm{I}(x,y) = \mathrm{I}(x,y) - \mathrm{I}(x,y-1). \tag{2.39}$$

These two derivatives can be connected according to a calculation rule like the mean value for the amount of the direction difference [4, 19]:

$$|\Delta_{xy}|_{\mathrm{m}} = \frac{|\Delta_x \mathrm{I}(x,y)| + |\Delta_y \mathrm{I}(x,y)|}{2}. \tag{2.40}$$

For $\Delta_x \mathrm{I}(x,y)$ and $\Delta_y \mathrm{I}(x,y)$ convolution matrices $K_x$ and $K_y$ can be written:

$$K_x = \begin{vmatrix} -1 & 1 \end{vmatrix}, \quad K_y = \begin{vmatrix} -1 \\ 1 \end{vmatrix}. \tag{2.41}$$

If the explained convolution procedure is applied, it is necessary to have a center in the matrix, which can not be found in the two matrices (2.41). This can be accomplished by inserting zero values into the matrices:

$$K_x = \begin{vmatrix} 1 & 0 & -1 \end{vmatrix}, \quad K_y = \begin{vmatrix} 1 \\ 0 \\ -1 \end{vmatrix}. \tag{2.42}$$

The Sobel operator can be used for the edge detection and uses convolution matrix with the following entries:

$$K_{Sx} = \begin{vmatrix} 1 & 0 & -1 \\ 2 & 0 & -2 \\ 1 & 0 & -1 \end{vmatrix}, \quad K_{Sy} = \begin{vmatrix} 1 & 2 & 1 \\ 0 & 0 & 0 \\ -1 & -2 & -1 \end{vmatrix}. \tag{2.43}$$

The Sobel operator is applied to an image that shows a doorplate in Figure 14.

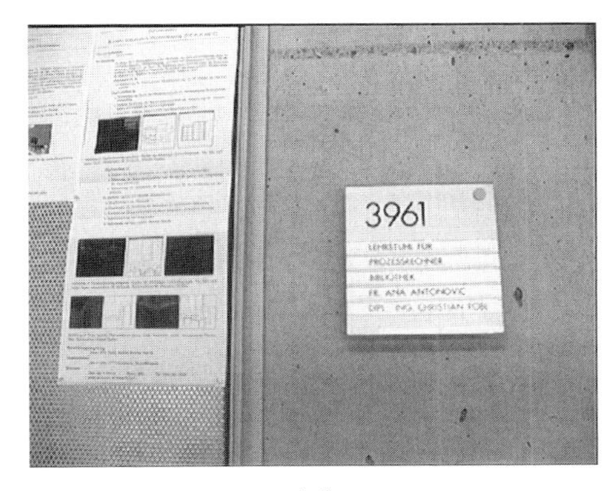


(a)

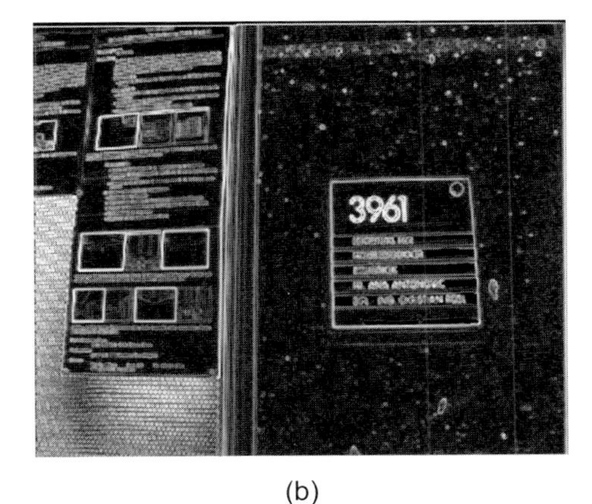


(b)

**Figure 14** Edge detection with the Sobel operator

The original image is shown in the left side of Figure 14. The result is shown in the right side of the image. The mask size of the Sobel filter used was $3 \times 3$ and the connection of the calculated values in $X$ and $Y$ directions happened with the calculation rule 'mean value for the amount of the direction difference'. If the doorplate is to be isolated with a segmentation process, the Sobel operator can be applied in any of the first steps to detect the edges in the image, and those that do not belong to the doorplate edges must be eliminated in further steps.

## 2.5 Skeleton Procedure

The calculation of the inner skeleton of an object can help to detect regions that belong to the object. The skeleton procedure is characterized by the following statements [3]:

1. Lines are not be interrupted or shortened.
2. The skeleton is created in the middle of the original region.
3. The lines have a width of one pixel.

Skeleton procedures are iterative. The processing time of these procedures increases strongly with the object size. The processing of the skeleton happens often by applying several operators. The object borders are removed one after another until only the inner skeleton of the object remains, which is represented by a line with the width of one pixel. An inner skeleton can also be calculated by the use of a so-called distance image of an object. The distance image is generated by the erosion of the object's foreground region until the whole region is eliminated. It is logged in every iteration step that pixels have been deleted. So the minimal distance to the object border is acquired for every pixel. The skeleton of the object can now be calculated by the examination of those pixels, which represent local maxima. These pixels will then be connected with lines that represent the object's inner skeleton [3].

## 2.6 The Segmentation of Image Regions

The segmentation of image regions can be executed top down or bottom up. The top-down approach needs properties for the decomposition of regions into segments. Here, it is, for instance, possible to use gray-value jumps, which indicate as a rule the boundary of an object. First, the bottom-up approach needs initial image segments of the region that must be determined. Therefore, region $A$ must be decomposed into segments $P_i$ $(1 \leq i \leq n)$ so that every pixel $p^j$ $(1 \leq j \leq m)$ of the region $A$ belongs exactly to one segment [3]:

$$\begin{aligned} &\forall p^j \in A : \exists k : p^j \in P_k \\ &P_j \cap P_l = 0 \; for \; j \neq l \\ &P_1 \cup \cdots P_n = A \end{aligned} \quad . \tag{2.44}$$

This problem can be solved in a trivial manner if every pixel represents one segment. This strategy provides too many segments, which are difficult to handle. Therefore, initial segments should be found preferably that do not enclose parts of different objects. These segments then become merged together by the use of a threshold value [3].

With region-growth procedures merging of pixels will be accomplished to one region providing a similarity criterion for the pixels is fulfilled. A region-growth procedure begins with start points and expands the regions from these points. The segmentation of image regions can be processed locally or globally. Pixels in a narrow neighborhood are inspected by the local segmentation and fused if they fulfill the similarity criterion. This procedure enables real-time operations. On the other hand, it often yields a merging of too many pixels because of the restricted area that is analyzed. Global processing uses larger neighborhoods and can provide better results. On the other hand, this strategy consumes more computing time. So real-time processing is not possible [3].

## 2.7 Threshold

Threshold operation [19, 20] is a simple but in many cases efficient method for segmentation. The threshold operator offers possibility to define a valid domain for gray values. Therefore, two gray values are necessary that are the lower boundary `min` and upper boundary `max` of the range:

```
threshold (InputImage, min, max, OutputImage) min <= max.
```

The threshold operator is applied to an input image. All gray values that have the value `min`, `max` or a value between `min` and `max` are chosen from the input image and are represented in the output image with the value one. The unselected pixels take the value zero. A formal description follows that describes the explanations:

$$\mathrm{I_T}(x, y) = \begin{cases} 1 \; \textit{if} \; T_{\min} \leq \mathrm{I}_G(x, y) \leq T_{\max} \\ 0 \; \textit{else} \end{cases} . \tag{2.45}$$

The usual threshold operator creates a binary image $\mathrm{I_T}$. The function $\mathrm{I_T}(x, y)$ examines the value 1 if the gray value $\mathrm{I}_G(x, y)$ is larger than or equal to $T_{\min}$ and smaller than or equal to $T_{\max}$.

Often the selected pixels are illustrated in the output image as white pixels and black pixels are used for the background. For example, if the input image is a gray-value image with maximal 256 gray values, the threshold operator can be used as shown in the following statement:

```
threshold (Doorplate, 0, 95, RawSegmentation).
```

The threshold operation is demonstrated with the gray-value image that was shown before in Figure 14. The input image contains dark text on the doorplate. This text will be segmented. Because the text on the doorplate is in black, the gray values of the text will be rather lower. Therefore, an acceptable domain is defined from 0 until 95 for the gray values, see Figure 15.

**Figure 15** The image `RawSegmentation`

The threshold operation provides the image `RawSegmentation` as the result in which the pixels that have gray values within the valid domain are illustrated as white pixels. All other pixels outside of the valid domain [0; 95] are represented as black pixels. In certain cases, like inhomogeneous illumination, the declaration of absolute values for the pixels, which will be selected, generates only unsatisfactory results. For example, if an image is taken from the doorplate in the early evening without artificial illumination, the gray values of the characters will be between 0 and 95 with high probability. But if the image is taken at midday, and the doorplate is strongly illuminated from the sun, it will probably be the case that some pixels of the characters have a value higher than 95 and therefore are not selected.

Some variants of the threshold operator exist to encounter such problems. They execute an automatic adjustment. It is not necessary to specify the valid domain for the gray values, because these operators use the gray-value distribution of the image as a guide [21]. Kittler *et al.* [22] have developed a threshold algorithm that executes an automatic threshold selection. For these purposes image statistics are used that do not require the computation of a histogram that represents the gray-level distribution.

The commercial image-processing library HALCON [4] offers a dynamic threshold operator:

```
dynamic_threshold (InputImage, SmoothedImage, OutputImage,
Offset, IlluminationOption).
```

An offset is available that can be controlled with a parameter. Very noisy regions are obtained usually if the offset matches to the interval [–1; 1]. An offset that is larger than 60 can have the effect that no pixels are selected. All pixels are probably selected if an offset is smaller than –60. The dynamic threshold operator receives two input images. One image is an original image. A smoothed image of the original image is generally used as the second image. The image $I_O$ will be the original image and image $I_S$ the smoothed image. As mentioned before, an offset *of* is also necessary [4]:

$$I_T(x, y) = \begin{cases} 1 & \textit{if } I_O(x, y) \leq I_S(x, y) + of \\ 0 & \textit{else} \end{cases} \quad . \tag{2.46}$$

$I_O(x, y)$ and $I_S(x, y)$ are functions that calculate gray values of the coordinates $x$ and $y$. The result of the dynamic threshold is the binary image $I_T$. Pixels are included in the binary image $I_T$ provided that they fulfill an inequality. The dynamic threshold operator can apply several inequalities. Which inequality holds is determined by the user with a parameter (IlluminationOption). The parameter can be initialized with values 'light', 'dark', 'equal', and 'not_equal'. If the value 'light' is selected, the following inequality is applied [4]:

$$I_O(x, y) \geq I_S(x, y) + of. \tag{2.47}$$

The option 'dark' effects the application of the formula

$$I_O(x, y) \leq I_S(x, y) - of, \tag{2.48}$$

the option 'equal' enforces the formula

$$I_S(x, y) - of \leq I_O(x, y) \leq I_S(x, y) + of, \tag{2.49}$$

and the option 'not_equal' the formula

$$I_S(x, y) - of > I_O(x, y) \vee I_O(x, y) > I_S(x, y) + of \tag{2.50}$$

[4].

Sahoo *et al.* [23] explain several variants of threshold operators and also report on evaluations of the variants performed with real-world images. Lee *et al.* [24] also executed evaluations for several threshold operators.

To improve the results of the threshold operator, it is often recommended first to apply a filter like a mean filter to smooth the gray-value distribution. The result of the filter can then be used as the input image for the gray values.

# 3 Navigation

Systems that control the navigation of a mobile robot are based on several paradigms.

Biologically motivated applications, for example, adopt the assumed behavior of animals [25, 26]. Geometric representations use geometrical elements like rectangles, polygons, and cylinders for the modeling of an environment [27, 28]. Also, systems for mobile robots exist that do not use a representation of their environment. The behavior [29] of the robot is determined by the sensor data actually taken [30]. Further approaches were introduced which use icons to represent the environment [31, 32].

## 3.1 Coordinate Systems

This chapter explains the use of coordinate systems in the robotics and conversions between these systems according to methods given in [33]. Movement in robotics is frequently considered as the local change of a rigid object in relation to another rigid object. Translation is the movement of all mass points of a rigid object with the same speed and direction on parallel tracks. If the mass points run along concentric tracks by revolving a rigid axis, it is a rotation. Every movement of an object can be described by declaration of the rotation and the translation. The Cartesian coordinate system is often used to measure the positions of the objects. The position of a coordinate system $\mathbf{X}_C$ relative to a reference coordinate system $\mathbf{X}_M$ is the origin $O$ from $\mathbf{X}_C$ written in coordinates from $\mathbf{X}_M$. For example, the origin of $\mathbf{X}_M$ could be the base of a robot and the origin from $\mathbf{X}_C$ could be a camera mounted on the robot. A vector of angles gives the orientation of a coordinate system $\mathbf{X}_C$ with respect to a coordinate system $\mathbf{X}_M$. By applying these angles to the coordinate system $\mathbf{X}_M$, it rotates so that it is commutated with the coordinate system $\mathbf{X}_C$. Angle $\alpha_C$ determines the rotation for the $X_M$-axis of $\mathbf{X}_M$, angle $\beta_C$ the rotation for the $Y_M$-axis of $\mathbf{X}_M$, and angle $\gamma_C$ the rotation for the $Z_M$-axis of $\mathbf{X}_M$. These angles must be applied to the original coordinate system of $\mathbf{X}_M$. The location of a coordinate system $\mathbf{X}_C$ comprises the position and the rotation in relation to a coordinate system $\mathbf{X}_M$. So the location is determined with vector $l_C$ that has six values

*Robot Vision: Video-based Indoor Exploration with Autonomous and Mobile Robots.* Stefan Florczyk

ISBN: 3-527-40544-5

$$l_C = (x_M, y_M, z_M, \alpha_C, \beta_C, \gamma_C). \quad (3.1)$$

The values $x_M$, $y_M$, and $z_M$ give the position in the reference coordinate system $\mathbf{X}_M$ and the angles $\alpha_C$, $\beta_C$, and $\gamma_C$ the orientation. It is possible to write the orientation of a coordinate system $\mathbf{X}_C$ in relation to a coordinate system $\mathbf{X}_M$ with the aid of a $3 \times 3$ rotation matrix. Rotation matrices consist of orthogonal unit vectors. It holds that:

$$M^{-1} = M^T. \quad (3.2)$$

The rotation matrix $M$ can be processed from elemental $3 \times 3$ rotary matrices $X_M(\alpha_C)$, $Y_M(\beta_C)$, and $Z_M(\gamma_C)$ of the three orientation angles $\alpha_C$, $\beta_C$, and $\gamma_C$. The rotation with $\alpha_C$ for the $X_M$-axis is stated as $X_M(\alpha_C)$, the rotation with $\beta_C$ for the $Y_M$-axis as $Y_M(\beta_C)$, and so forth.

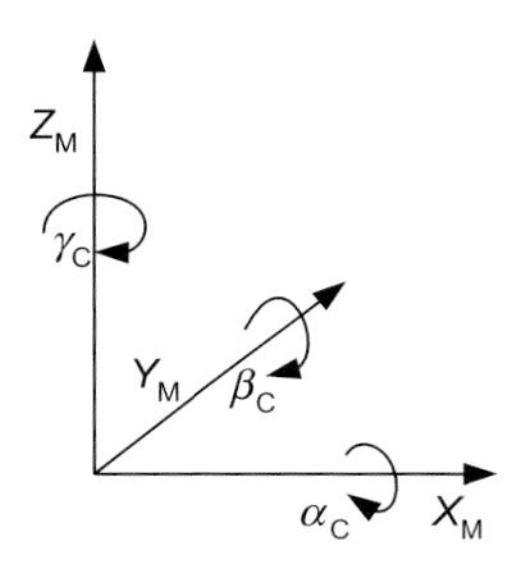


**Figure 16** The six degrees of freedom [33]

Figure 16 shows the three axes $X_M$, $Y_M$, and $Z_M$ for the coordinate system $\mathbf{X}_M$. Rotation angles $\alpha_C$, $\beta_C$, and $\gamma_C$ are attached to the axes. The reference coordinate system $\mathbf{X}_M$ can be moved in the direction of the three axes to obtain the coordinate system $\mathbf{X}_C$. It can also be rotated around the three axes. This means that six degrees of freedom are possible.

Homogeneous transformation [34, 35] uses a $4 \times 4$ matrix for rotation and translation. The transformation of the coordinate system $\mathbf{X}_M$ into the coordinate system $\mathbf{X}_C$ is written with the homogeneous matrix $H_{M,C}$. Let $(x, y, z)_C$ and $(x, y, z)_M$ be the position of the same scene point in homogeneous coordinates [36], then the following formula holds:

$$(x, y, z)_C = H_{M,C} \cdot (x, y, z)_M \quad (3.3)$$

and always

$$H_{M,C} = (H_{C,M})^{-1}. \quad (3.4)$$

The location of a rigid object in the coordinate system $\mathbf{X}_C$ and in the coordinate system $\mathbf{X}_M$ can be represented with a homogeneous matrix $H_{M,C}$:

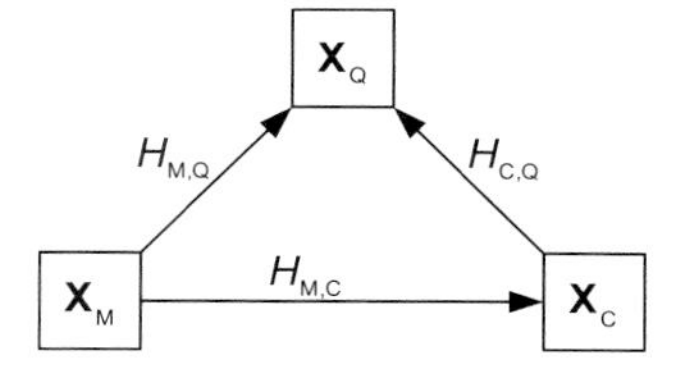


**Figure 17** Conversion from locations [33]

A further coordinate system $\mathbf{X}_Q$ is introduced in Figure 17. If relations are given as in Figure 17, $H_{M,Q}$ can be processed:

$$H_{M,Q} = H_{M,C} \cdot H_{C,Q}. \tag{3.5}$$

Often several coordinate systems are necessary in robotics. For example, the view of a robot is represented with coordinate system $\mathbf{X}_M$. Therefore, the origin of $\mathbf{X}_M$ is the base of the robot. If the robot is equipped with a sensor like a camera, it can be used as a second coordinate system $\mathbf{X}_C$, whereby the origin of $\mathbf{X}_C$ represents the camera that is mounted on the robot, see Figure 18.

**Figure 18** Coordinate systems for a mobile robot

For example, the mounted camera can be used for depth estimation. The taking of two images from different positions can perform this. It is possible to process the depth for the taken scenario with the coordinates of the camera's two different positions.

## 3.2
## Representation Forms

### 3.2.1
### Grid-based Maps

Grid-based maps use a coordinate system like the polar coordinate system to represent the environment. Egocentric maps, whose creation was inspired by biological research [37], are a form of grid-based maps that generate the model of the environment from the viewpoint of the robot, based on its coordinate system. Egocentric maps can be developed at runtime to recognize objects, which are dynamically changing in the environment like people, vases of flowers, and so forth, for which a persistent representation is impossible. For example, equidistant occupancy grids can be used for these purposes. So egocentric maps are, for example, helpful for collision avoidance. Because of the dynamic aspect of these egocentric maps, frequent updates are required that are often calculated several times per second by the evaluation of information that is provided from sensors like sonar, a ring bumper, or a laser. Allocentric maps are independent of the view of the robot and can be derived by merging several egocentric maps [38], see Figure 19.

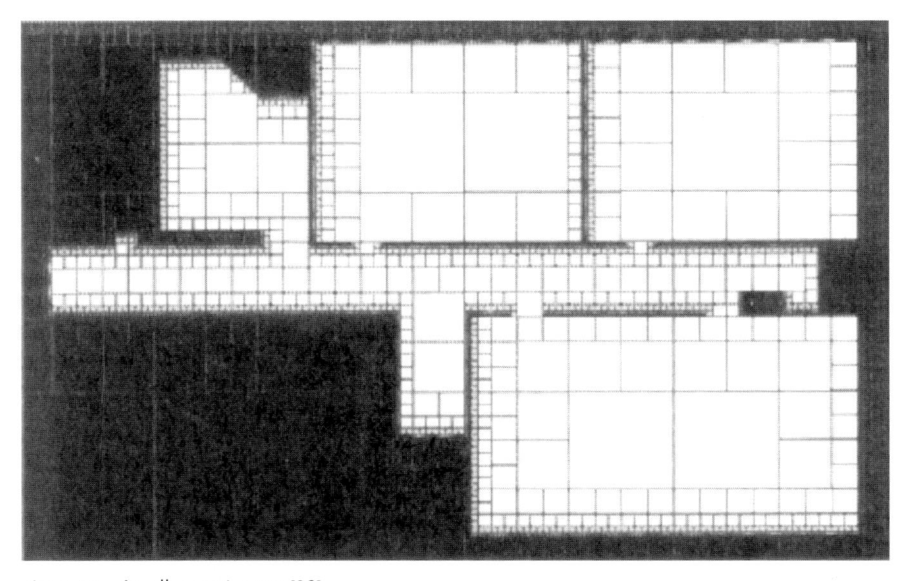

**Figure 19** An allocentric map [38]

The more static character of allocentric maps can result in more persistent maps with a probabilistic representation of occupancy [38]. Robot RHINO's navigation system creates and uses these maps [39]. The self-localization in an allocentric map can, for instance, happen with the Monte Carlo localization. The equidistant probabilistic occupancy grids are, analogous to egocentric maps, often chosen. Many realistic applications require maps to represent areas of approximate $1000\,m^2$ and more. Sonar was mainly used as the distance sensor in the past. The reduced costs for lasers offer the possibility to produce grid maps with higher resolution. These two aspects can yield allocentric maps, which should not be stored in the entire RAM (random access memory), because this can result in long processing for algorithms like the path planning. A solution to handle this problem could be to decompose the entire map into smaller portions so that only a part of the map is stored in the RAM and the rest remains only in the permanent store. If the robot solves tasks that are formulated by humans it can be necessary to use additional knowledge about objects, which can be stored in a knowledge database. This knowledge is attached to the objects represented in the allocentric occupancy maps. With allocentric occupancy maps it is difficult to handle problems if an object covers many cells in the grid map, because cells are independent and do not share information in these maps. To attach information to a region that consists of several cells, region maps are used. For example, a table can consist of many cells that are linked to such a region and attached with information that every location in this region belongs to the table. Further region maps provide the advantage that a quick update in some dynamic cases can be performed. If the robot detects an unequivocally identified object at an unexpected position, it is able to free the region from the former occupancy of that object. When the robot knows about the dynamic behavior of some objects, like animals, it is possible to predict the temporal occupancy of a region [38].

### 3.2.2
### Graph-based Maps

Topological maps can be used to solve abstract tasks, for example, to go and retrieve objects whose positions are not exactly known because the locations of the objects are often changed. Topological maps are graphs whose nodes represent static objects like rooms, doors, and so forth [38], see Figure 20.

The edges between the nodes denote 'is part' relationships between the objects. For example, an abstract task formulated by a human user could be to fetch a wrench of size 13. To solve this task, the robot proceeds with the aid of the topological map along an appropriate path. Therefore, abstract information like 'wrenches are in the workshop' is sufficient in the knowledge base to enter the approximate position of the wrench, whereby the topological map contains a node that represents the workshop. For example, it is possible to derive a topological map from an occupancy map by a segmentation process to obtain objects, which can be represented as nodes. Relationships must be analyzed between these objects to obtain edges [38].

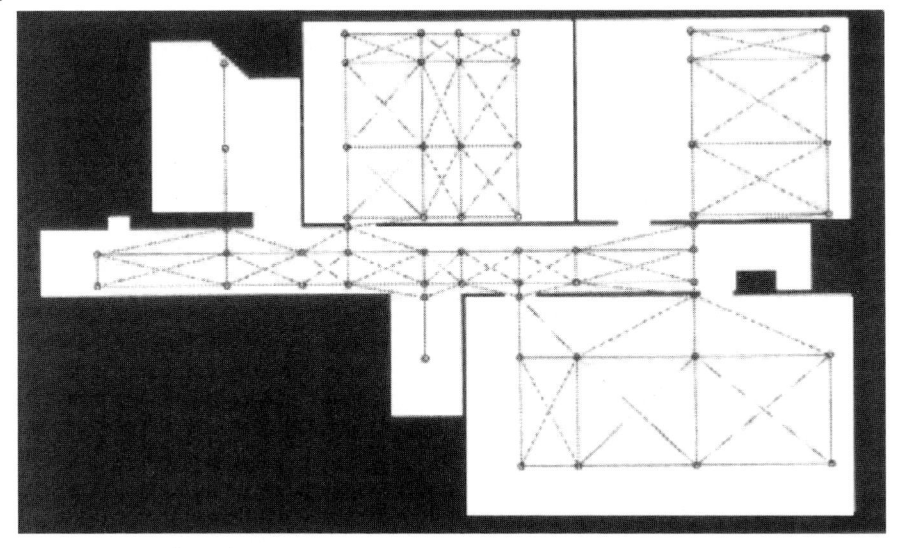

**Figure 20** A topological map [38]

## 3.3
## Path Planning

Several approaches for path planning exist for mobile robots, whose suitability depends on a particular problem in an application. For example, behavior-based reactive methods are a good choice for robust collision avoidance [38].

Path planning in spatial representation often requires the integration of several approaches. This can provide efficient, accurate, and consistent navigation of a mobile robot and was, for example, shown by Thrun and Bücken [40], who combined topological and occupancy maps for path planning.

### 3.3.1
### Topological Path Planning

Topological path planning is useful for the creation of long-distance paths, which support the navigation for solving a task. Therefore, those nodes representing, for example, free region space are extracted from a topological map, which connect a start point with a target point. The start point is mostly the actual position of the robot. To generate the path, several sophisticated and classical algorithms exist that are based on graph theory, like the algorithm of the shortest path [41].

To give best support for the path planning, it could be helpful to use different abstraction levels for topological maps. For example, if the robot enters a particular room of an employee for postal delivery, the robot must use a topological map that contains the doors of an office building and the room numbers. On the other hand, if the office building consists of several areas, and the robot has to go to one of the areas, it is sufficient for the robot to use a topological map that represents only the different areas without details such as office rooms and so forth. The possibility to use topological maps with different abstraction levels helps to save processing time. The static aspect of topological maps enables rather the creation of paths without information that is relevant at runtime. The created schedule, which is based on a topological map, holds nothing about humans or animals, which occupy the path. In that case it is not possible to perform the schedule. To get further actual information, the schedule should be enriched by the use of more up-to-date plans like egocentric maps [38].

### 3.3.2 Behavior-based Path Execution

Behavior-based path execution can be used for collision avoidance, position monitoring, and goal achievement. These types are performed partially concurrently. If the robot expects after its current position the goal region, two scenarios can occur. The robot reaches the goal region. In that case the robot can switch from position monitoring to goal achievement. If the robot does not enter the expected goal, it can use strategies to solve the problem, which depends on an event occurring. If a human obstructs a path to a goal region, the robot can ask him to release the path. Because of necessary collision avoidance it can happen that a calculated path is useless. This requires the processing of a new schedule [38].

### 3.3.3 Global Path Planning

Schölkopf and Mallot developed a view graph [42] to represent a scenario with several distributed obstacles. Simple collision avoidance was implemented. Data from an infrared sensor were evaluated. The collision avoidance started if an object was over one centimeter away. If the distance of a robot was less than one centimeter, the robot has first to back up and must then flee the obstacle. The description of the scenario happens with visual input, which provides gray values of a surrounding panorama. The visual input of one situation in an environment is called a 'snapshot' and is stored as a node in the view graph. So a node represents a sensorial perception at a specific location in the environment. Snapshots do not contain metric information like distances or angles, only local views. Edges denote spatial connections between snapshots and are used if the moving direction from one snapshot to another can be processed. For these purposes a so-called homing algorithm [43] is taken. If a robot is located at a snapshot, the system must choose a direction. Therefore, the directions to all snapshots in the neighborhood are estimated and the angles between the snapshots are determined. The largest angle provides the new direction for the exploration. The direction is chosen in such a way that the robot drives in the middle of the region between the two snapshots for which the largest angle was calculated. Snapshots can not be taken if the collision avoidance runs. Therefore, it is not possible to get snapshots in the neighborhood of obstacles. Connections between snapshots in different graphs are calculated with a homing algorithm. These determined connections model the spatial relationships in the environment. The processing of the connections between snapshots occurs under the assumption that visible landmarks have a constant distance to a snapshot. Therefore, the approach is not appropriate if many obstacles exist with different distances to the snapshot [44].

### 3.3.4
### Local Path Planning

Tani [45] has developed an approach for local navigation that first creates a topological map of an environment. This happens in the learning phase. If a goal is to be reached, local path planning is performed under the condition that the length of run is minimal. A behavior-based robot detects unobstructed areas. The centers of the detected areas are used for the path planning. If several areas are available that are examined from sensory data the system uses a decision process to select between the alternatives. A mobile robot was used that provided sensorial data from a laser range finder.

### 3.3.5
### The Combination of Global and Local Path Planning

Täubig and Heinze [46] described an approach that integrates local and global navigation for a robot in an office building. Typically corridors, office rooms, and crossways exist in an office building or similar building. To explore an unknown environment, the robot drives very fast without collisions in corridors. If the robot arrives at a crossway or an office room, then a decision situation emerges in which the robot will choose a direction that results in an unexamined region. A direction that would end in a known blind alley should be avoided. Two levels can be derived from this description. The robot will drive fast without collision at the local level. The level is active if only one direction exists in which the robot can drive. On the other hand, the global level is chosen if the robot is in a state in which it can select several directions.

The drawing in the top-left part of Figure 21 shows a situation in which the robot drives along a corridor. The robot is not in a decision situation and the local level is therefore active. In the top-right part of the figure, the robot is in a corridor in a decision situation, because it has reached a crossway. The local level detects four directions that can be chosen. The selection of one direction happens with the global level. After the selection, the robot drives again along a corridor, which means that the robot navigates in the local level. This is shown in the lower-left part of the figure. The drawing in the lower-right part shows that the robot is driving on a free-space area. The local level offers several directions. The global level selects one of these directions.

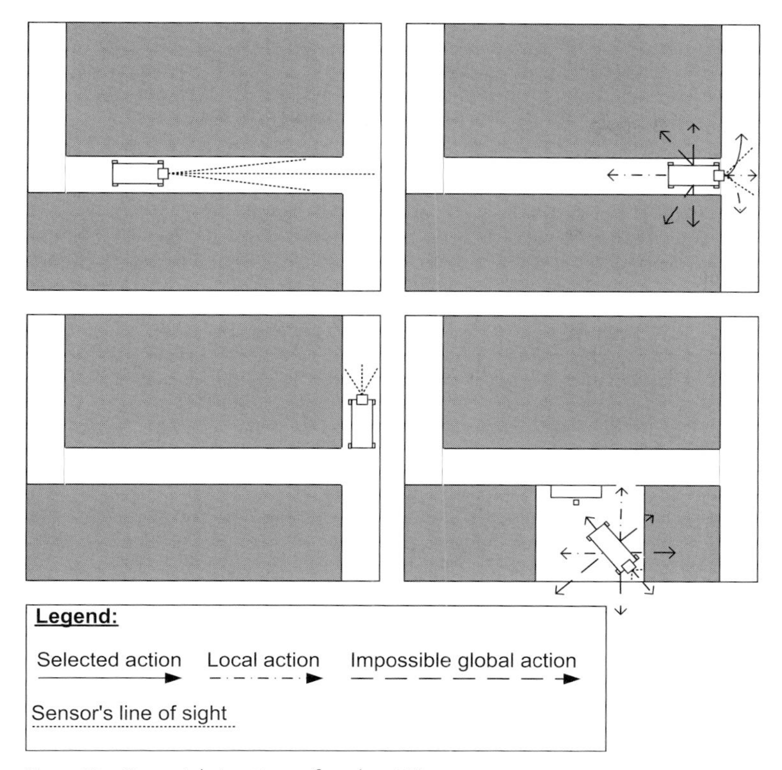


**Figure 21** Sensorial situations of a robot [46]

The view graph is expanded with local edges. A local edge connects two snapshots. This means that the connected snapshots can be reached among one another with local navigation. If the robot is not in a decision situation, the local navigation is used until a decision situation emerges. Then the sensorial input provides information for the new snapshot. The path that was followed by the local navigation is represented with a local edge, which ends in a node that represents the new snapshot. The robot has to choose the new direction in a decision situation. Therefore, for every snapshot that is connected to the actual snapshot it is the size of the unexplored region with a global algorithm processed that does not consider obstacles. So the local level is used to get the size of the area that is occupied by obstacles. This size is then subtracted from the size of the region processed with the global algorithm. The biggest remaining unexplored region is chosen. The center of the chosen region provides the next location for the exploration of the next direction. The robot is moved in the examined moving direction until a new sensorial input is found that is dissimilar to the last snapshot. Then a global edge is inserted into the view graph. This sensorial input constructs the new snapshot and is inserted into the graph and connected with the former snapshot by a global edge. If the processed snapshot is similar to a snapshot that is stored in the graph, the robot tries to reach this snapshot. In this case the movement of the robot is the basis for the global edge.

An example of a view graph with global and local edges is shown in Figure 22.

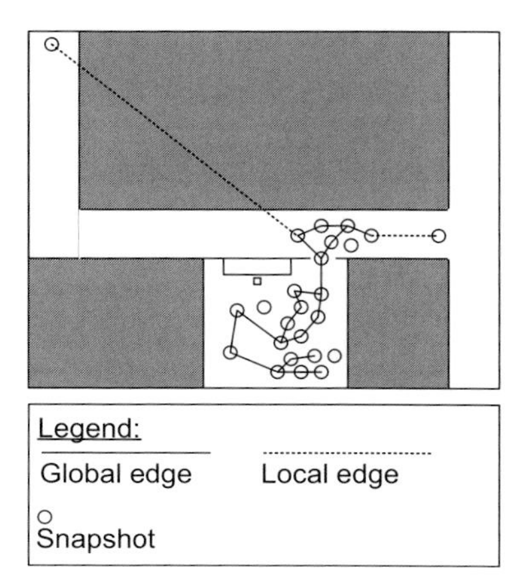


**Figure 22** Example of a view graph with global and local edges [46]

In the top-left part of Figure 22 the start location can be found for the view graph. A local edge was inserted into the graph, because there was no decision situation. A local edge is denoted with an interrupted line. A decision situation is found at the office door. So a snapshot is modeled with a node. The entire decision situation is constructed with several nodes that are connected with global edges. The accumulation of nodes can also be observed in the room. Local edges can be found in corridors.

## 3.4 The Architecture of a Multilevel Map Representation

Figure 23 shows an example of a multilevel representation.

Necessary information is collected with sensors like cameras, laser scanners, and so forth. After the interpretation, the information can be used to construct an egocentric map that shows a snapshot of an actual situation and supports the robot by the localization. Because of the allocentric map's more permanent character, it is necessary to use the egocentric map for an update of the allocentric map. It is possible to derive regions that contain relevant objects from the allocentric map with the aid of segmentation and classification. These regions with the attached objects are represented in the region map and yield the foundation for the abstracter topological map that is connected with the knowledge base, which holds information about the detected objects. The knowledge in connection with the topological map enables the robot to perform the task planning. During the execution of the schedule, the robot is able to use modules for collision avoidance, position monitoring, and goal achievement by controlling the actuation. During the runtime of the robot these calculations are repeated several times, so it is possible to ensure consistency between the maps with different abstraction levels. Of course it is possible that inconsistency occurs temporarily between the different maps. This requires then perhaps that a new schedule must be created [38].

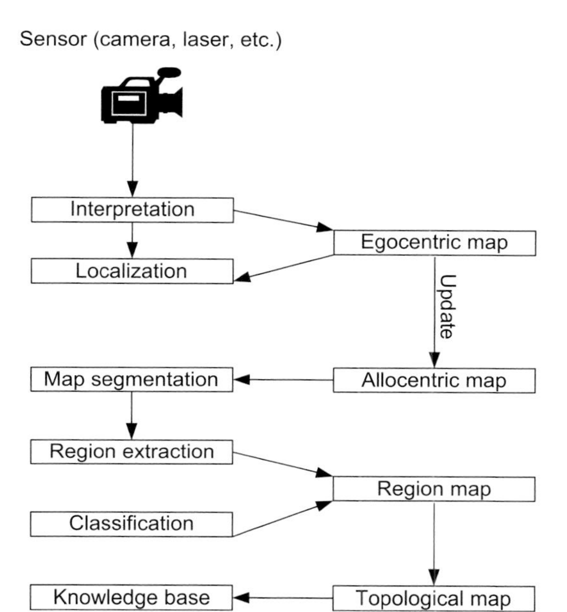


**Figure 23** The architecture of a multilevel map representation [38]

## 3.5 Self-localization

Stochastic methods like Markov localization and Monte Carlo localization are used to determine the actual position of a mobile and autonomous robot.

Burgard *et al.* [1] described extended Markov localization, which is a component of robot RHINO that was deployed for 6 days as a museum tour guide. The extended Markov localization revealed its robustness during the test. The localization was successful although most of the robot's sensors were blocked by people who followed the robot. The extended Markov localization eliminated damaged sensor data.

Monte Carlo localization provides a multimodal distribution for position estimation. Several local maximums exist. Every maximum represents a possible position of the robot. The Monte Carlo localization applies a cyclic computation to eliminate local maximums until only one maximum remains. As long as more than one maximum exists, the position is ambiguous. The dynamic aspect of the environment makes it difficult to find the correct position. Small objects like vases and animals often change position. The brightness is changing so that, for example, a reference image can provide a different middle-gray value in comparison to the image actually taken. The Monte Carlo localization uses a distribution of samples in the state space. Every sample represents a position. For these purposes three-dimensional Cartesian coordinates can be used. Each sample is associated with a probability that represents the possibility that the corresponding sample is the correct actual position of the robot. The probability is derived from reference information like an image of the position. The initialization is performed in the first step of the Monte Carlo localization. The position is unknown, and the samples are distributed in the state space. Reference information is used to determine a probability for every sample. To obtain good results, the determination of the probability is supported by an error measure. For example, brightness differences between the actual position and the reference position are compared. The robot performs a position alteration in the second step. Hence, the samples get a new state, because their positions and orientations change. The reference information can be newly processed with the aid of interpolation and subsequently used to get new probabilities for the samples. The cumulative probability over all samples is then standardized to one and the new probabilities multiplied by the former probabilities. Once again the cumulative probability is standardized to one. Samples that have a low probability, are eliminated in the last step. The resulting freed samples are now distributed in regions with higher probabilities. Steps two and three are iterated until samples remain only in the true position of the robot. The procedure of the Monte Carlo localization is represented in Figure 24 [47].

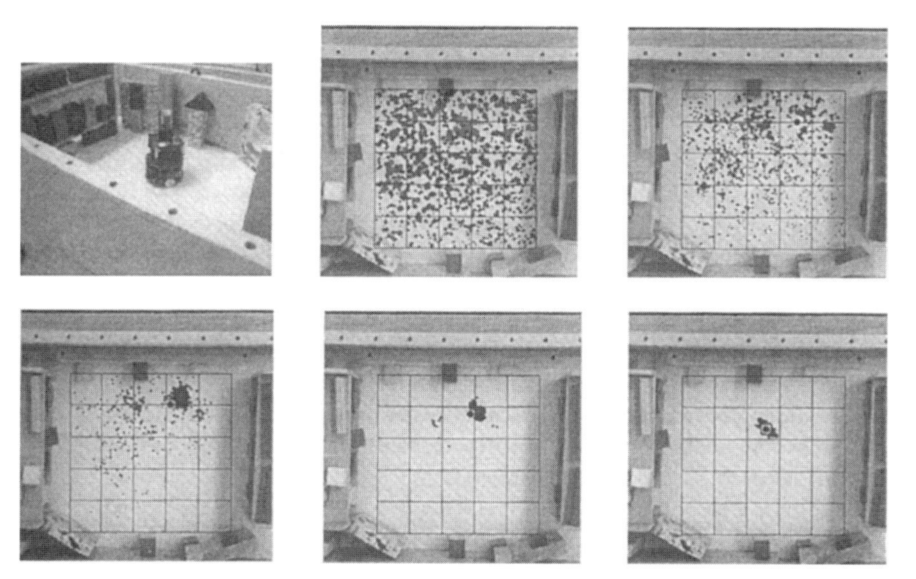

**Figure 24** An example of the Monte Carlo localization [47]

The top-left picture shows a robot in a test environment. In the top-middle picture the test environment is shown with the initial state of the sample distributions in a grid. A second step was executed, with the result that the samples are clustered in some regions. This is shown in the top-right picture. The lower pictures show the distribution after some repeats of steps two and three. It can be observed that one position remains that represents the true position of the robot [47]. Also, systems have been successfully developed that are able to perform the localization and mapping for mobile robots concurrently [48].

# 4
# Vision Systems

## 4.1
## The Human Visual Apparatus

### 4.1.1
### The Functionality

A rather abstract description of the human visual apparatus is now given. Consider Figure 25.

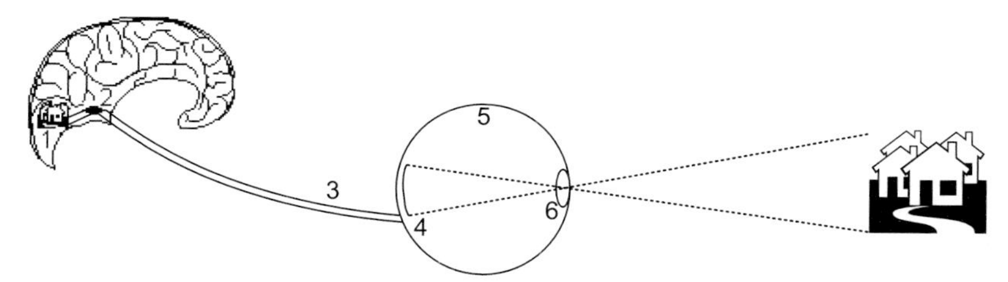


**Figure 25** An abstract view of the human visual apparatus [49]

The light incidence is controlled by an opening, which belongs to the iris (5). The iris is a muscle and it is able to alter the size of its opening by contraction. This means that the amount of the light incidence, which strikes the retina (4), becomes greater or smaller. So the retina has an image of the scene that was captured by the eye. The focusing of the retinal image is executed by altering the curvature of the lens (6), which is behind the opening of the iris. The retinal image is converted into electrical signals by photosensitive cells on the retina [49].

The density of the cells in the center of the retina is the highest. This means that the resolving power is best here. The resolution decreases in the direction towards the edge of the retina. The cells build squares. These squares are responsible for strengthening the contrast of the image by using complementary colors and can be grouped into two kinds. One of the two squares is responsible for bright colors and the other for darker colors. The squares in the center of the retina enclose many

*Robot Vision: Video-based Indoor Exploration with Autonomous and Mobile Robots.* Stefan Florczyk

ISBN: 3-527-40544-5

photosensitive cells in comparison to those squares that belong to the edge of the retina. Therefore, the retina has a higher resolution in the center [49].

The photosensitive cells exist in two kinds as mentioned before: cones are responsible for the determination of the gray values and three types of rods for the recognition of the colors red, green and blue (RGB). The cones have a higher photosensitivity than the rods. The color signals of the rods are compounded after being generating and then transmitted by the visual nerve (3) and the lateral geniculate bodies (2) to the visual cortex (1). The lateral geniculate bodies are responsible for stereo vision. The signals are stored in Luv-channels. The L-channel involves information on the brightness and the uv-channels on the colors [49].

### 4.1.2
### The Visual Cortex

Figure 26 shows five of the six layers of the visual cortex responsible for vision. The deepest layer six projects back to the lateral geniculate bodies and is not illustrated in the figure [50].

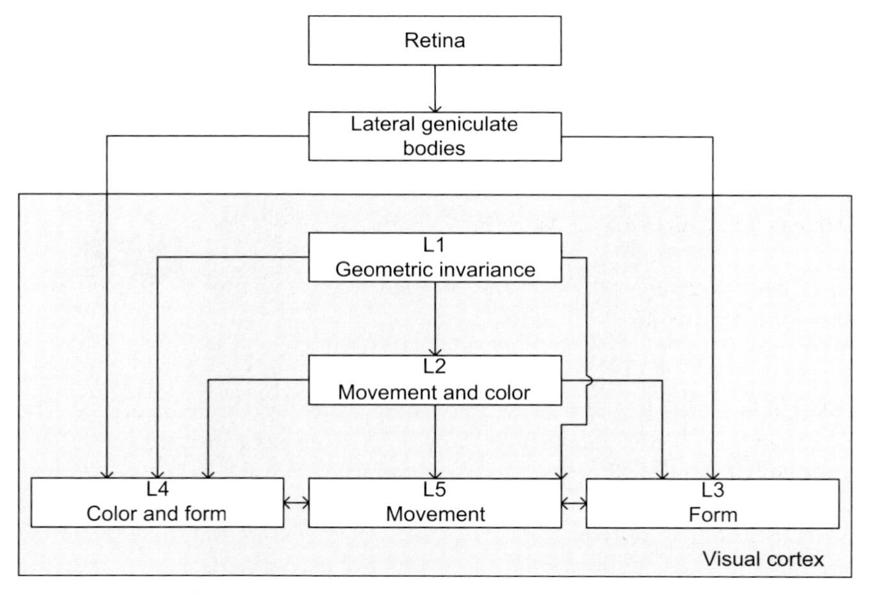


**Figure 26** Layers of the visual cortex [49]

The first layer (L1) processes invariant images. Information about color, movement, and form is preprocessed in the second layer (L2). The processing of the information takes place in parallel in the layers three (L3), four (L4) and five (L5). Hence, the visual cortex is also responsible for the preprocessing and processing of the image. It receives the signals from the lateral geniculate bodies. First, the signals are transferred to the first area. The invariant coding of the image is executed here. Invariant coding means that an image is processed here, which is a prototype without loss of image information for the set of all shifted, size-altered, and rotated images. The representation of the retina's visual information in the visual cortex's first layer happens by the so-called logarithmic polar transformation [49].

The cortical columns exist in the first layer. The layers of the retina belonging together are reflected here. Architectural layers, for instance, are responsible for the recognition of colors and movement. Rotations are handled by orientation columns. The projection of the retinal image happens in a manner such that the center of the retinal image provides a greater part to the projection in the first area than the regions belonging to the border of the retinal image. So the resolution power of the retina is found again in the visual cortex. A position-invariant projection of the retinal image occurs by the displacement of the regarded object into the image center. The position-invariant projection of the image is a prerequisite for the preparation of the invariant image concerning the invariance of the image dimensions and invariance of the image rotation. An invariant image with respect to the image dimensions is represented in the visual cortex by activating different cortical columns. The size alteration of the retinal-image effects a shifting in the visual cortex. Images in the retina of different sizes are represented in the visual cortex as identical images in different cortical columns. The presentation of rotated images on the retina takes place in an analogous way in the visual cortex. In this case varied orientation columns are activated [49].

## 4.2 The Human Visual Apparatus as Model for Technical Vision Systems

The parts of a computer-vision system work in a way similar to the human visual apparatus, see Figure 27.

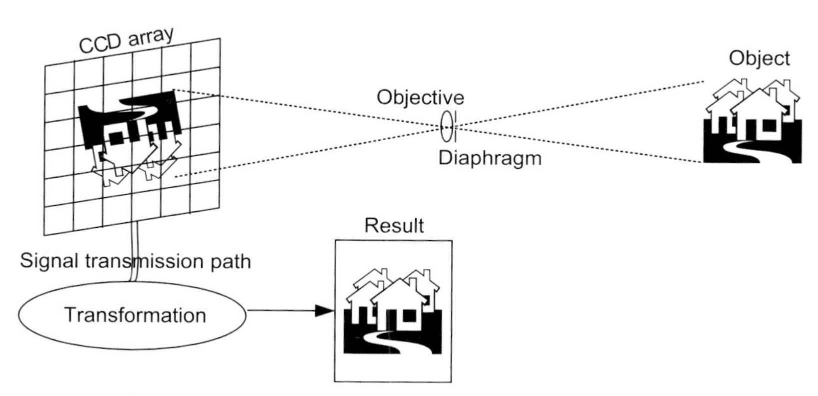


**Figure 27** Abstract presentation of a technical vision system [49]

The diaphragm of an objective controls the light incidence and corresponds to the iris. The objective is responsible for focusing and can be compared with the lens and the CCD[1] array is oriented at the retina. The transmission of the generated electrical signals takes place with the signal transmission path that is comparable with the visual nerve. Transformation algorithms adopt the invariant coding of the visual cortex. The logarithmic polar transformation is realized [51–53] in some technical systems. The logarithmic polar transformation can only be performed if the image is invariant coded [49].

### 4.2.1 Attention Control

Area, which is necessary for problem solving, can be determined with attention control. Features like the color, contrast, or movement can be used for attention control to provide the area. If the human visual apparatus is considered, the attention is also controlled besides these features by the resolving power of the sensor. The retina has the highest resolution in the center, so the attention is directed to objects that are represented in the center rather than at the border of the retina. It could also be shown that the attention control by the human visual apparatus is strongly affected by the problem that should be solved. The advantage of the human visual apparatus has been adopted for technical vision systems. The visual search or object detection finds instances from objects by often using attributes like the color [54].

Burt uses a window with a higher resolution than the entire representation of the scene. The window is moved over the entire scene representation and conducts the attention to the relevant scene detail [55].

An analogous approach is proposed by [56]. The entire scene is analyzed by operators to detect stored object models in knowledge base. If the scene possesses objects that have a resemblance to the object models, but by the resolution of the whole scene the equality can not unequivocally prove that it is performed by an operator that is responsible for detecting such conflicts, then this part of the scene is examined with a higher resolution. Krotkov [57] combines a stereo vision system and a focusing approach and uses the visual attention control to obtain a partial reconstruction. In the first step the stereo vision system is used to generate a rather coarse reconstruction. Points in the scene are determined by the use of the coarse reconstruction, which are then used for a more precise reconstruction with the focusing approach. The system from Abbott and Ahuja [58] develops global depth maps by stepwise expansion to represent the scene successively more completely. If a depth map is generated, the borders of the map are examined to obtain the expansion of the global depth map. New views must be positioned at the examined borders. The next view must be determined with a cost function. Views that are nearby the scene are preferred, because this helps to minimize the danger that remoter objects could occlude nearer objects [54].

1) charge coupled device

### 4.2.2
### Passive Vision

Effort has been made in the past to develop a theory for machine vision that describes how technical systems can solve tasks similarly as they would be tackled by the biological vision apparatus. Marr [59] has developed such a theory. He described the information processing of image data. This approach is regarded as passive vision, because the scene is observed from a passive motionless observer [60].

Horn [61] proposed that the entire scene must be represented by symbols in the process of passive vision. This should happen with the intention that all machine-vision tasks can then be solved easily. Marr's theory consists of three parts. First, there is the processing theory. The aim of the problem will be analyzed here. Once this has happened, it should be possible to provide necessary requirements so that the machine-vision task can be solved. The second part (representation and algorithm) describes the process of input processing and output. The last part (hardware) yields explanations for the implementation of algorithms in hardware [60].

### 4.2.3
### Active Vision

Meanwhile, the opinion exists that the complete reconstruction of the scene is not necessary to solve specific machine-vision problems. In active vision the technical observer interacts with its environment and performs only selected vision similar to the biological vision apparatus. Biological vision always follows a determined purpose and is therefore only a part of the complete vision. For example, the human visual apparatus is restricted to a specific wavelength band of the light. So a general vision is not possible. Presently, computer vision is strongly restricted by assumptions, for example, a required illumination. If the necessary prerequisites are not fulfilled, the robustness of a computer-vision application rapidly diminishes. Often, a total outfall can be observed. So algorithms are required that provide a robust recognition. Problem solving should happen by using relaxed presumptions wherever possible [60].

Active vision can be further specified. Controlling the zoom, the position of the camera, the diaphragm, and the line of sight can actively influence the taking of images. Subject to a problem, image regions can be classified into different importance levels. This will enable saving of processing time. The interpretation of the taken images can also be seen as a part of the active vision and is called active symbolic vision. A further form of active vision is stimulated active vision. The behavior patterns of an observer are examined here to obtain relevant information. For example, to separate an object from its background, the object can be fixed by the observer. This has the effect that the object becomes sharply in contrast to the background. Active and exploratory perception is a further attempt to give an explanation for active vision. To survive in its environment, a biological system is performing strongly selective vision by the exploration of the surrounding. For these purposes mobility is a necessary prerequisite. The environment is also changing during the time interval. So the system must be able to recognize this by learning. The set of information that is absorbed is too large for a precise analysis. This means that the information must be filtered. The system is equipped with some procedures to acquire information for a precise selection. The choice of an appropriate procedure is executed by using a cost estimate. The utilization of a procedure must result in lower costs in comparison to benefit that can be measured by the information gain [60].

### 4.2.4
### Space-variant Active Vision

Space-variant active vision is a newer form of active vision. The technical vision apparatus is based on the vision apparatus of upper vertebrates. In the upper vertebrates the resolving power is highest in the center of the retina as described earlier. So the sensor surface will also have the highest resolution in the center. This means that it is only necessary to use a window with a high resolving power in the center of the sensor. Nevertheless, the resolving power in the direction towards the border of the sensor remains high enough to control the attention and to conduct the center to the important regions. So the selection that is an essential attribute of active vision is already considered by the construction of the sensor and results in a strong reduction of the data set and therefore to less processing time. But in this case it is a problem as standard algorithms of the image processing can not be used [60].

The resolving power is constant over the entire sensor surface and cameras are only static in the case of space-invariant passive vision systems. This is approximately comparable with the vision apparatus of the goldfish and requires a large increase in the number of pixels by the magnification of the sensor's $\Xi/K$-quality ($\Upsilon$). $\Xi$ indicates the field of vision and $K$ the maximal resolution power. If $\Upsilon$ is doubled, it is necessary to quadruple the number of pixels. So this approach does not support the selection and provides no contribution to save processing time. Today, most image-processing systems have an architecture that follows the approach of space-invariant passive vision systems, because it is simpler to construct and supports standard algorithms. The space-invariant active vision supports a homogeneous sensor with an active vision apparatus like a camera that can be rotated and inclined. For example, a camera that has the flare angle of 50°, which is expanded by an actuation to 150°, provides a $\Xi/K$ quality improved by a factor of 9. The architecture today is often used in technical vision systems. In contrast, the space-variant passive vision is used less often and combines a space-variant sensor and a rigid camera. For example, a camera with a resolution of $2000 \times 2000$ pixels was used. The area of $512 \times 512$ was moved over the camera image. So it was possible to imitate the camera movement. But prices for cameras with such a resolving power are high and also require processing time and memory [60].

## 4.3 Camera Types

### 4.3.1 Video Cameras

Video cameras transmit analog signals. These signals are converted with a frame grabber into digital images with the typical size of $768 \times 512$ pixels which need 0.44 MB. Video cameras are widespread in computer vision, because they are relatively cheap and appropriate for real-time purposes. The drawback of video cameras is the rather low resolution [62].

### 4.3.2 CCD Sensors

It is possible to convert a conventional photographic camera into a digital camera with CCD sensors by the mounting of the CCD sensor into the image plane. Additionally, a hardware device like a PCMCIA card is necessary to store the data. In contrast to conventional cameras, the images taken can be viewed on a computer directly, because development time is not necessary. This can improve the quality of the pictures, because poor quality pictures can be replaced on the spot. The resolution differs and has typically $2000 \times 3000$ pixels. An image with this resolution consumes 6 MB in the case of a gray image and 18 MB if it is a color image. The size of a CCD sensor is about $2.4 \times 1.6\,\text{cm}^2$ [62].

A pinhole camera is a very simple model for a CCD camera. The hole is the origin of camera coordinate system $\mathbf{X}_\text{C}$. Figure 28 shows the model of the pinhole camera [33].

The camera's optical axis is represented by the $Z$-axis of the camera coordinate system $\mathbf{X}_\text{C}$ with axes $X_\text{C}$, $Y_\text{C}$, and $Z_\text{C}$ and origin $O_\text{C}$. A three-dimensional scene point $X$ is projected from the hole onto the CCD array $C$. The CCD array is symbolized with the two-dimensional sensor coordinate system $\mathbf{X}_\text{S}$ with axes $X_\text{S}$, $Y_\text{S}$, and origin $O_\text{S}$. $O_\text{S}$ is determined by the point of intersection between the optical axis and the sensor $C$. Focal length $b$ is determined by the distance between the origin $O_\text{C}$ of the camera coordinate system $\mathbf{X}_\text{C}$ and the origin $O_\text{S}$ of the sensor coordinate system $\mathbf{X}_\text{S}$. The principal axis distance [64] is another name for the focal length that is sometimes chosen. A point $(x, y, z)_\text{C}$ in the camera coordinate system $\mathbf{X}_\text{C}$ is mapped to the sensor coordinate system $\mathbf{X}_\text{S}$ with the following equation [33, 63]:

$$\begin{bmatrix} x_\text{S} \\ y_\text{S} \end{bmatrix} = \frac{b}{z_\text{C}} \cdot \begin{bmatrix} x_\text{C} \\ y_\text{C} \end{bmatrix}. \tag{4.1}$$

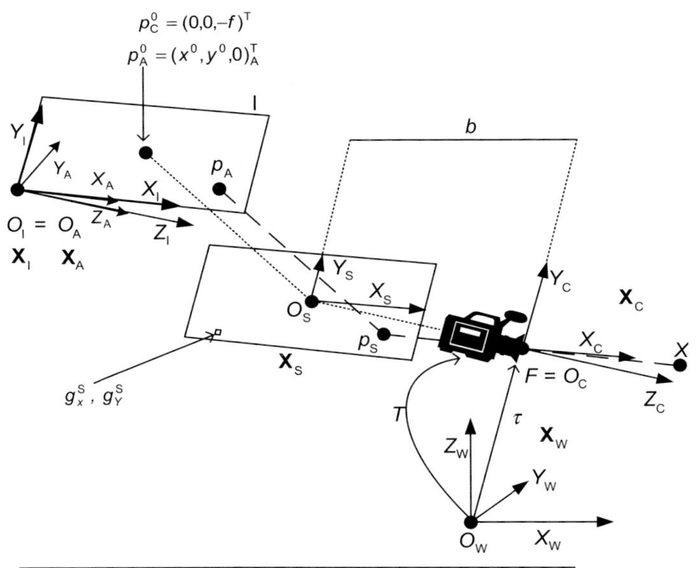


**Legend:**

| | | | |
|---|---|---|---|
| Image Euclidean coordinate system with axes $X_I$, $Y_I$, $Z_I$ | $\mathbf{X}_I$ | Focal length | $b$ |
| Image affine coordinate system with axes $X_A$, $Y_A$, $Z_A$ | $\mathbf{X}_A$ | Image plane | I |
| Sensor coordinate system with axes $X_S$, $Y_S$ | $\mathbf{X}_S$ | Focal point | $F$ |
| Camera Euclidean coordinate system with axes $X_C$, $Y_C$, $Z_C$ | $\mathbf{X}_C$ | Origin | $O$ |
| World Euclidean coordinate system with axes $X_W$, $Y_W$, $Z_W$ | $\mathbf{X}_W$ | Scene point | $X$ |
| CCD array | $C$ | Two-dimensional points | $p_A$ $p_S$ |
| Principal points | $p_C^0$ $p_A^0$ | Translation | $\tau$ |
| Size of a sensor pixel | $g_x^S$, $g_y^S$ | Rotation | $T$ |

**Figure 28** The pinhole camera model [33, 63]

The mapping between the sensor coordinate system $\mathbf{X}_S$ and the image affine coordinate system $\mathbf{X}_A$ with axes $X_A$, $Y_A$, $Z_A$ and origin $O_A$ is performed by a frame grabber. The result is the projected point $p_A$. $p_A^0 = (x^0, y^0, 0)_A^T$ is the principal point of $\mathbf{X}_A$. $\mathbf{X}_C$ has coordinates $(0,\ 0,\ -f)^T$ for the principal point now denoted with $p_C^0$ to distinguish it from the coordinates belonging to $p_A^0$. It can be seen that the point of intersection between the optical axis and the digital image determines the principal point. The axes $X_I$, $Y_I$, $Z_I$ of the image Euclidean coordinate system $\mathbf{X}_I$ with origin $O_I$ are aligned with the axes of the camera coordinate system $\mathbf{X}_C$. $X_I$ and $Z_I$ are also aligned with the axes $X_A$ and $Z_A$ of the image affine coordinate system $\mathbf{X}_A$. $Y_A$ can have another orientation as $Y_I$. $g_x^S$ and $g_y^S$ are scaling factors and model the size of a sensor pixel. The coordinates $(x, y)_A$ of the image affine coordinate system $\mathbf{X}_A$ have no dimensions. The unit is a pixel. The unit of the scaling factors is a meter. The origin $O_A$ is the upper-left corner of the digital image. The processing of the pixel coordinates for a digital image follows [33, 63]:

$$x_A = \frac{x_S}{g_x^S} + u^0, \quad y_A = \frac{y_S}{g_y^S} + v^0. \tag{4.2}$$

The camera coordinate system $\mathbf{X}_C$ can be derived by applying rotation $T$ and translation $\tau$ to the world coordinate system $\mathbf{X}_W$ with axes $X_W$, $Y_W$, $Z_W$, and origin $O_W$ [63].

### 4.3.3 Analog Metric Cameras

Analog metric cameras are often used for taking aerial images to get a high resolution. Therefore, the analog image of a metric camera is analyzed offline with a scanner. The image is scanned with a high resolution $(g_x, g_y)$. This strategy is very time consuming and requires much memory. For example, a gray image with the resolution of $16\,000 \times 16\,000$ pixels takes 256 MB. Analog metric cameras in conjunction with an offline scanning process are applied by the taking of aerial images, because these images often have the size of $23 \times 23\,\text{cm}^2$ [62].

The projection is now sketched. Analog metric cameras can be represented with a pinhole camera just as the CCD camera. Figure 29 shows a model of the pinhole camera that takes an aerial view [65].

The image plane has the distance $c$ to the projection center $F$, in the camera model, whereby $c$ is a camera constant that is used in photogrammetry instead of the focal length. The camera constant $c$ includes additionally linear radial distortions. The geometrical record axis stands vertically on the image plane and cuts the image plane in the principal point. The offset between the principal point and the image center is represented with the principal point offset $h$. Clockwise Cartesian coordinate systems are used to describe the projection between a scene point $X$ and a pixel $p_S$. A scene is represented with the rigid world coordinate system $\mathbf{X}_W$ with axes $X_W, Y_W, Z_W$. To measure the camera view, the camera coordinate system $\mathbf{X}_C$ with axes $X_C, Y_C, Z_C$ is used. $Z_C$ stands vertically on the image plane. The image plane is spanned by $X_C$ and $Y_C$. The projection between the spatial point $X$ and the projection center $F$ runs along the camera's line of sight $A$ [65].

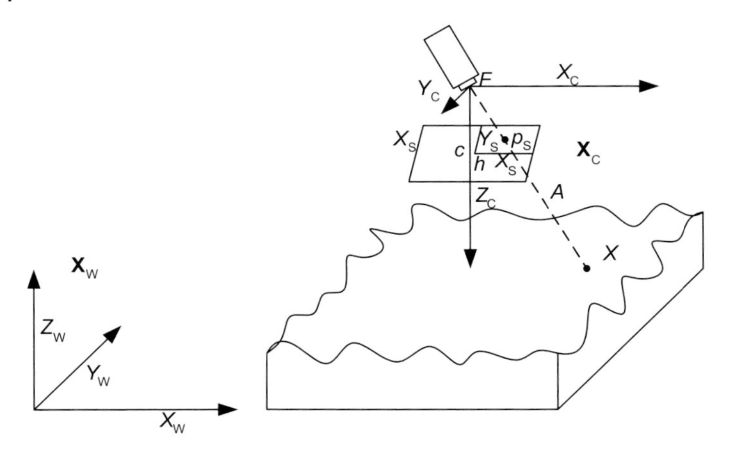


**Figure 29** Model of a pinhole camera recording an aerial view [65]

In the $Fz_C HL$ system from Yakimovsky and Cunningham the camera constant $c$ is considered additionally in vectors $L$ and $H$ [66]. In this case the image plane has the distance of one to the origin of the $Fz_C HL$ system. The principal point offset is represented with $HLz_C$ [65]:

$$p_S = \frac{1}{(X-F)^T \cdot z_C} \cdot \begin{pmatrix} (X-F)^T \cdot H \\ (X-F)^T \cdot L \end{pmatrix} + h \tag{4.3}$$

$$\text{with:} \;\; H = \frac{c}{g_x} x_C + h_x \cdot z_C \,, \quad L = \frac{c}{g_y} y_C + h_y \cdot z_C. \tag{4.4}$$

# 5
# CAD

## 5.1
## Constructive Solid Geometry

Constructive solid geometry (CSG) is a volume-oriented model. Objects are represented with base elements or primitives (*BE*) like a square stone, cylinder, sphere, cone, and so forth. The approach uses association instructions to link the base elements to new more complex elements (*CE*). A binary tree is used to represent the CSG. The leaves of the tree are used to represent the base elements. The association of the elements happens with association operators (*AO*) like unification, intersection, difference, and complement, which are known from the set theory. These association operators are applied to base elements and more complex elements, respectively. The result is a more complex element [67], see Figure 30.

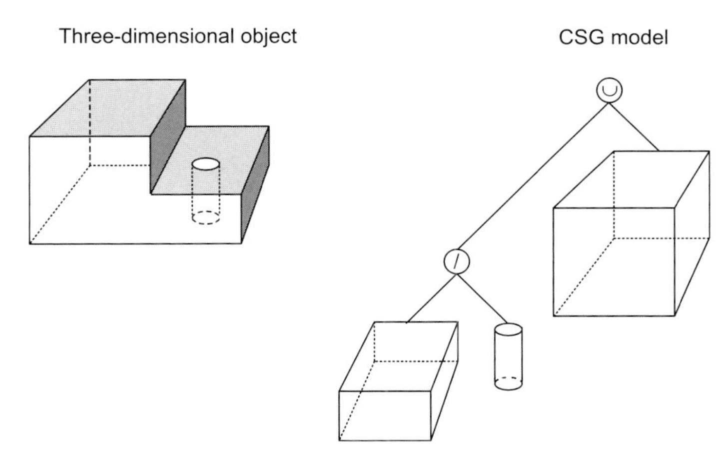


**Figure 30** Representation of a three-dimensional model with CSG model [68]

The formal description that depicts the functioning of the association operators is:

*Robot Vision: Video-based Indoor Exploration with Autonomous and Mobile Robots.* Stefan Florczyk

ISBN: 3-527-40544-5

$$SE = \{BE, CE\}, \tag{5.1}$$

$$AO = \{\cup, \cap, \ldots, \}, \tag{5.2}$$

$$ao_j(se_1, \ldots, se_n) \rightarrow CE \qquad se_i \in SE, i = 1, 2, \ldots, n; ao_j \in AO,$$

$$j = 1, 2, \ldots m. \tag{5.3}$$

Let $SE$ be a set that contains base elements $BE$ and complex elements $CE$. Set $AO$ contains association operators like unification, intersection, and so forth. Operator $ao_j$ needs as input a subset $(se_1, \ldots, se_n)$ of $SE$. Base operators and more complex operators can be contained in the subset. The utilization of the operator $ao_j$ provides, as a result, a more complex element $CE$.

The more complex elements are attached to the nodes of the binary tree. If a base element is used in a model, the size and location of the base element must be kept in the model. The set of base elements is changeable. This means that it is possible to add or delete base elements. If such a process has taken place, it is necessary to calculate a new binary tree. Therefore, the calculation must start by that branch of the tree where the alteration has been performed. It can be rapidly processed with a binary tree whether a spatial point is covered by an element or not. The CSG model does not possess information about surfaces and edges. If visual presentation is required, it is necessary to process the information by using stored data. This has the advantage that storage can be saved. Otherwise more processing time is required for visualization. Every alteration of the line of sight requires a new processing of the surfaces and edges [67].

## 5.2 Boundary-representation Schema (B-rep)

Object boundaries are pictured in a model with boundary-representation schema. Volumes are modeled with surface elements, whereby a surface element is described with its borderlines. A borderline can be reconstructed if the end points are stored [67], see Figure 31.

A more complex calculation is necessary in comparison to the CSG model to decide if a spatial point is covered from an object or not. The visualization of B-rep models is simply possible, because the entire required information is stored. So the alteration of the line of sight does not much affect processing. Hence, the need for memory increases in comparison to CSG. The B-rep model is appropriate to represent freeform. With CSG this is difficult, because freeform can not be modeled sufficiently with geometric base elements [67].

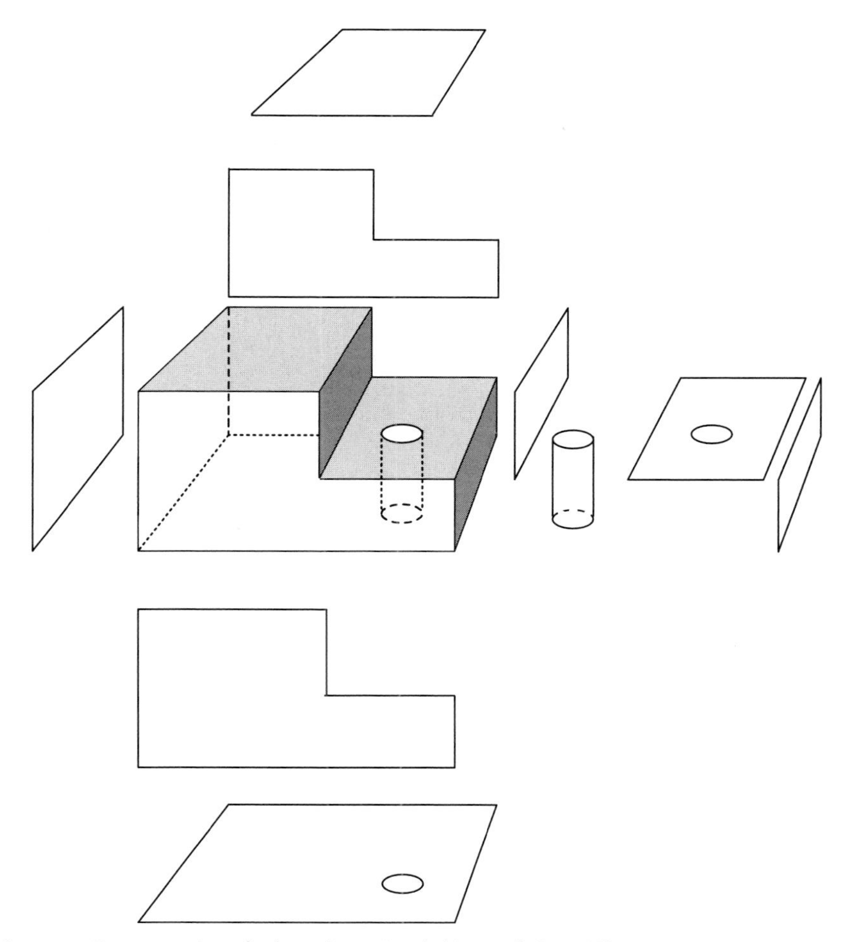

**Figure 31** Representation of a three-dimensional object with B-rep [68]

## 5.3 Approximate Models

Approximate models are an alternative approach to geometric models. Functionality follows the divide and conquer principle. This principle decomposes a problem into parts. Every part can then be analyzed separately from the entire problem so that a problem reduction is performed. The partial solutions are merged and the whole solution built. The binary-cell model is a further term for approximate models, because the Euclidean space is represented with three-dimensional cells. Trees are used to represent hierarchical approximate models [67].

### 5.3.1
### Octrees

The quadtree model was developed in image processing to represent two-dimensional objects. Octrees are an advancement of quadtrees to model three-dimensional objects. The objects are approximately modeled with cubes in octrees [69]. A cube is represented with a node in the tree and can have one of three possible types [70], see Figure 32.

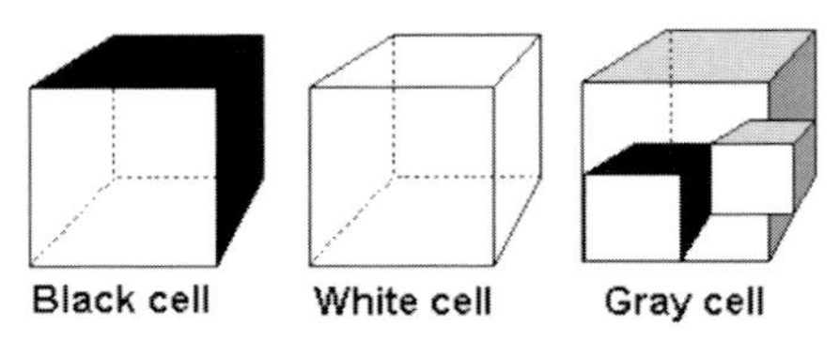


**Figure 32** Three types in the octree [67]

If the cube is completely within the object a black cell is used. The opposite case holds if the cube is completely outside the object. These two types are end nodes in the tree and not further refined. A gray cell is used if one part of the cube is covered by the object and the other not. This gray cell is refined. The refinement is executed with the bisection of the three Euclidean axes, which are used to model the coordinates of the cube. The bisection affects eight new cubes and the procedure starts again and can therefore be programmed with a recursive procedure. The procedure stops if an abort criterion, like a required solution, is fulfilled. At the moment it must be decided whether the remaining gray cells are related to the object or the background. Therefore, the object is modeled by the black cells and the corresponding gray cells [67].

### 5.3.2
### Extended Octrees

Extended octrees comprise the three types of the octrees and have further three additional types [71], see Figure 33.

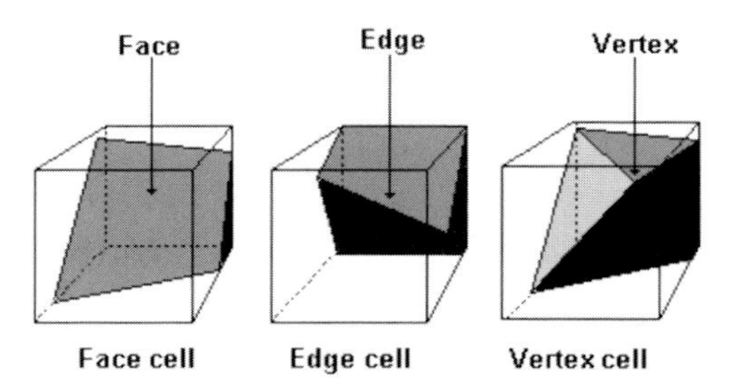


**Figure 33** Additional types in extended octrees [72]

The face cell is used if the cube contains a piece of the object's surface. The edge cell represents a part of an edge and the two accompanying neighboring surfaces. The third type is the corner cell that comprises an object vertex and those edges that run into the corner [71, 73].

### 5.3.3 Voxel Model

Voxels are volume elements that are used to represent objects. A voxel in a model can belong to an object or not. Therefore, a voxel model can only be an approximation of the modeled object [74, 75]. The maximal resolution is controlled by the size of the voxel [76]. To get a more exact modeling with voxel models, they are often arranged in octrees [77]. The precision of an object model can also be improved with a higher granulation, but this requires an exponential increase of the data set. Three kinds of voxel models are in use [67], see Figure 34.

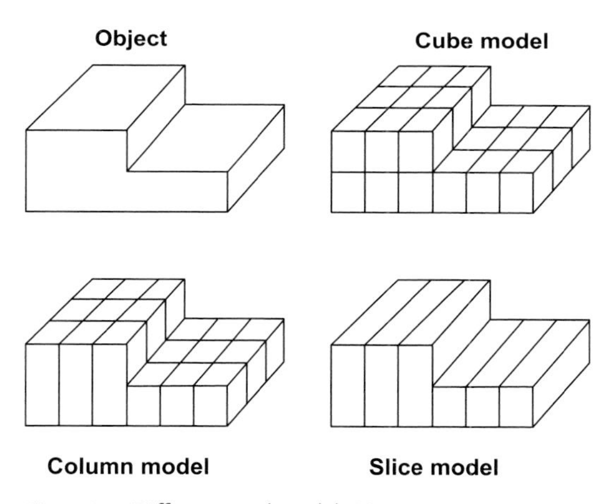


**Figure 34** Different voxel models [67]

The cube model consists of cubes that all have the same size. The column model uses square stones that all have the same width and length but different heights. The third model is the slice model. An object is represented with slices of the same thickness. The slices can be described with their edges or with their surfaces. The precision of the object representation increases with decreasing slice thickness, but then also needs more memory [67].

## 5.4
## Hybrid Models

As discussed earlier every model has its advantages and drawbacks. Two or more approaches can be combined to use the advantages of different models like the CSG and B-rep. Much calculating effort for visualization is needed if the CSG model is used when the line of sight has changed. When a CSG model is used as a primary model, which means that the modifications are executed in the CSG model, the B-rep model is additionally used for the purpose of visualization. Hence, a B-rep model is attached to every CSG node in the tree. The reverse case is also used. In this case B-rep is the primary model and completed with the CSG model [67].

## 5.5
## Procedures to Convert the Models

Several approaches were suggested to convert a B-rep model to an octree approximation.

The approach of Tamminen and Samet that consists of two phases performs the conversion successively. It analyzes the surfaces of an object separately. The result of this procedure provides approximate surfaces that are connected together. In the second step the nodes of the octree are classified with colors like black, white, and gray. The examination of a node's neighborhood is performed for these purposes to get necessary information for the classification and to determine the connected regions [78].

The approach of Kela converts B-rep into an octree in four steps. The algorithm operates recursively and decomposes, at each recursion level, the gained gray cells into eight successors. The entire model is initially contained in one gray cell. The partition of the vertices is executed in the first step. In the second step the partition of the edges takes place, and the partition of the surfaces happens in the third step. The classification of the cells with the accompanying surfaces and edges is done in the fourth and last step [79].

Kunii *et al.* [80] use four phases to convert octrees into B-rep models. First, an octree is transformed into an extended octree whose entities are labeled in the second phase. Tables are derived in the third phase that contain B-rep information. The fourth phase creates a sequence of Euler operations that describe the created B-rep model.

Conversion from B-rep to extended octrees can also be realized. The approach from Carlbom *et al.* [72] uses two steps to gain a result. The surfaces of the object are converted step by step into the tree in the first step. The cube is divided into eight derived cubes. This affects cuttings in the surface representation, which means that a pseudogeometry with pseudoedges and pseudovertices exists. It must be decided in the second step whether the node of the tree represents a part within the object or outside of the object. This examination happens with an algorithm.

Also, a reconversion from an extended octree to B-rep can be done in two steps. First, only the vertex cells are inspected and listed. For every face that belongs to a detected vertex, its two accompanying faces at the vertex are registered. This information is necessary to detect polygons in the second step [71].

A conversion from B-rep to CSG is difficult, because ambiguities can exist. The approach of Vossler uses a sweep procedure to generate simple objects, which are recognized with techniques known from pattern recognition. Simple objects can be combined into more complex objects [81]. Objects that are generated with the sweep procedure are also called production models. Models whose creation is performed with the Cartesian product, which is applied to geometric elements, belong to the production models. Rotation models, translation models, and trajectory models are such production models [82].

## 5.6 The Use of CAD in Computer Vision

Several approaches have been reported that combine CAD and computer vision.

ACRONYM is an early implementation that offers the user the possibility to create volume models of three-dimensional objects. The user also provides spatial relationships of the objects and subclass relationships. The created volume models and further data acquired from the user are then used for object recognition in images with a computer-vision system [83]. But existing CAD models, which were originally developed for product modeling, can also be used to extract features from them. The obtained data can then be used for recognition tasks executed by a computer-vision system [84, 85]. Flynn and Jain propose interpretation tables that are generated offline from CAD models. The interpretation tables are used to detect correspondences between primitive scene surfaces and primitive model surfaces [86]. Surface reconstruction can be based on stochastic computations to eliminate noise, blurring, nonlinearity, and discontinuity [87–89].

To realize a computer-vision application that is based on CAD, an approach can be used in which the model knowledge is represented in a CAD database. The CAD database is connected to the computer-vision system. Of course, different data types between the computer-vision system and the CAD database can occur. This requires type conversions between the CAD data types and the data types of the computer-vision application that can be performed by some systems automatically [90].

Horaud and Bolles use models of objects that are to be detected in a muddle of several objects. Therefore, the object to be detected can be partially occluded. To guarantee a reliable and fast recognition under these conditions, an incremental approach is used that first tries a recognition by the use of only one object feature. Only when a robust detection can not be based on one feature is a second feature taken. The number of object features is increased in a recognition process until a certain recognition can be guaranteed [91].

The approach of Wang and Iyengar uses the B-rep model in three-dimensional recognition and position determination from images. Area segments are used as feature primitives. These feature primitives are represented by translation and rotation invariance. Local curvature attributes are therefore used [92].

It can be an advantage to use a CAD model for a geometric representation of the objects within a computer-vision system to avoid type conversions. Furthermore, it is possible to use well-known CAD techniques for the modeling and the calculation of the objects in a computer-vision system. But the use of CAD in computer-vision systems must be reflected carefully. Models in CAD systems are complete. This enables a graphical representation. The necessary information is read from a CAD database or is derived from several views, which are integrated. A single two- or three-dimensional image often exists only in computer-vision applications that do not show an entire object in the scene. In this case it is not possible to obtain the entire view of an object. A reference coordinate system for an object is used in CAD models. The coordinate system is object centered, which means that the origin of the coordinate system refers to the object center. Computer-vision systems use a reference coordinate system too, but this is observer centered. For example, this can be the camera. In this case it is necessary to determine the position and the orientation of an object in a first step to generate a connection between the two coordinate systems. The view of an object is gained from a model in graphical applications. A computer-vision application has an inverse strategy. First, graphical representations are taken from which data and finally a model are derived. For example, this can be a B-rep model. It was shown that the advantages of a B-rep model for representation are also valid for the inverse case in computer vision if the objects are described by their contours and area segments. This requires uniform data structures in computer-vision applications, which comprise the following features [93]:

1. The data structures should contain representations from feature primitives, which can be extracted from model data and also from image data.
2. The representation of the feature primitives must be based on local attributes of an object and it is necessary that the representation is rotation and translation invariant.
3. Geometrical relations between feature primitives must be describable explicitly.
4. A complete representation must be designed that can be used in a CAD model. The usage of this representation in a computer-vision system must be viewed as pseudo-CAD that refers to the incomplete scene from which the CAD model is derived.

An example of the usage of a CAD model in computer vision is now explained in the following paragraphs according to the explanations in [93].

### 5.6.1 The Approximation of the Object Contour

The representation of objects occurs with a B-rep model in computer-vision systems. Several attributes like the contour, length of the contour, and so forth were used. A Boolean variable was used to show if a contour was closed or not. The geometrical description used an object-centered coordinate system. The pseudo-B-rep representation was generated from sensor data that used a camera-centered coordinate system. A polygon approximation was used for the reconstruction. The polygon approximation describes the object contour $CO$ with a sequence of corners:

$$CO = \{(x, y, z)^{i}_{C} | i = 1, 2, \ldots, n\}. \tag{5.4}$$

Parameter $lc$ denotes the edge length of the contour $CO$. Conversion from the three-dimensional representation into the two-dimensional representation was performed. Processing time should be reduced with this policy. Therefore, it holds that $z^{i}_{C} = 0$ for all corners in the new model. Also, parameters exist in the B-rep data structure, which are used to execute the transformation between the camera-centered coordinate system and the contour-centered coordinate system. It is possible to generate a transformation matrix between model and camera coordinate system with these parameters. So the object's position can be determined. Closed curves can be written with $\Theta(lc)$ function for shapes. A starting point must be determined on the shape's contour. The $\Theta(lc)$ function must be defined that measures angles in the shape's boundary as a function of arc length $lc$ [94]. An example of a possible $\Theta(lc)$ function is:

$$\Theta(lc) = \cos\left(\frac{y^{i}_{C}}{x^{i}_{C}}\right), \quad i = 1, 2, \ldots, n. \tag{5.5}$$

In Figure 35 a polygon is illustrated that can be constructed with an appropriate $\Theta(lc)$ function.

The figure shows that the steps in the $\Theta(lc)$ function correspond to the polygon's corners. The angles and lengths of the polygon areas can also be found in the B-rep representation. So processing time can be saved, because the $\Theta(lc)$ function does not have to be newly processed in further steps. The rotation of a contour is reflected by the shifting of the $\Theta(lc)$ function along the $\Theta$-axis.

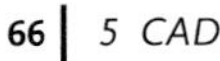

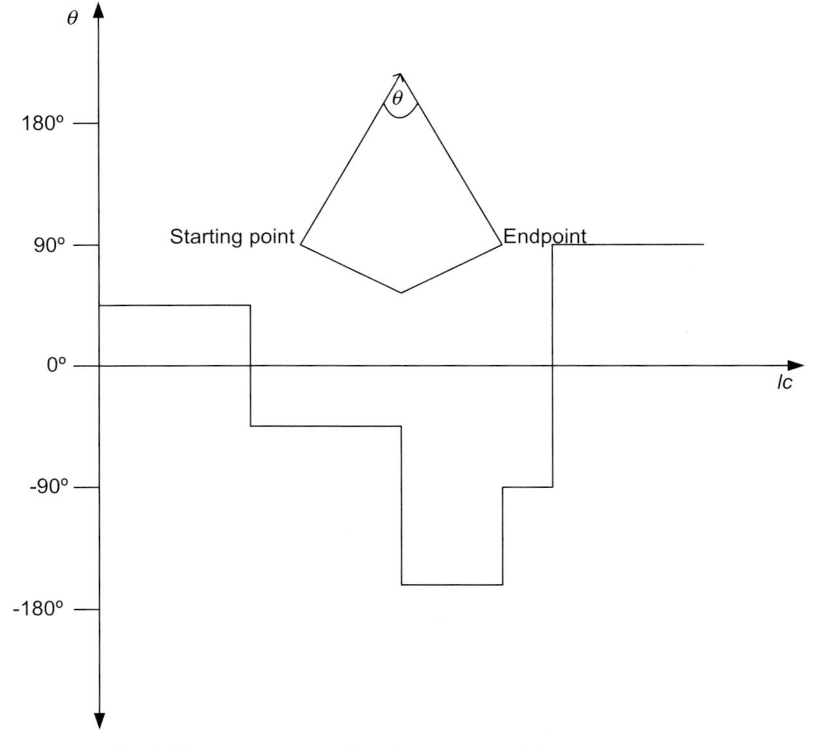


**Figure 35** The $\Theta(lc)$ representation of a polygon contour [93]

5.6.2
## Cluster Search in Transformation Space with Adaptive Subdivision

The coherent subsets of an entire set's elements can be found with the cluster search in transformation space. The elucidation follows explanations in [93]. A parameter vector is used for the description of the elements. Additionally, weighting factors can be used for the description. The weighting factor can serve as the quality factor for the particular element. Those elements that belong to the same partial set are characterized by qualitatively similar parameters. This can be geometrically observed by a cluster that is constituted from the particular elements in the transformation space. The cluster is valued with an entire quality, for example, this can be the sum of the element's single qualities. It is the cluster of all possible clusters chosen by which the entire quality is the highest. The functionality of the cluster search is now shown by the description of the algorithm. The examination of an optimal subset uses defined uncertainty areas and local qualities. An optimal subset is characterized by the intersection of uncertain areas, where the sum of the accompanying single qualities is maximal.

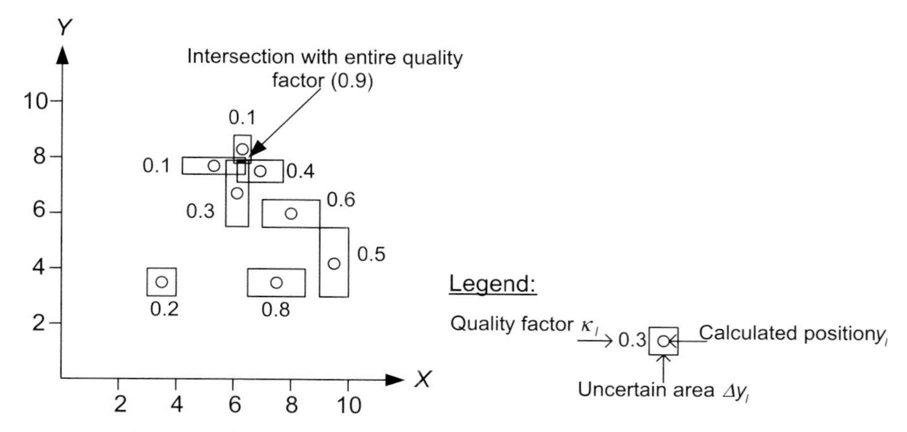


**Figure 36** Cluster search in a two-dimensional transformation space [93]

Figure 36 shows a two-dimensional transformation space. The two-dimensional coordinates of the elements represent the calculated position in the transformation space. In the figure it can also be seen that every element has a quality factor and a defined uncertain area. The entire quality factor is calculated by the addition of the single quality factors belonging to those elements by which the defined uncertain areas have a common intersection. The algorithm that processes an optimal subset by progressive subdivision of the transformation space is shown in Figure 37. The description of the algorithm uses C++ style. The kernel of this implementation suggestion is enclosed in two methods, which are part of the class 'intersection_detection'. The class uses further classes and accompanying members whose description is not provided.

```
constraints intersection_detection::best_subset_search(int
d_max, double q_min)
{
   q_best = q_min;       //Private member
   best_box = 0; //Private member
   /* 'all_candidates' is a private member of class 'box' that
   contains all candidates enclosed in transformation space and
   'con' is a private member of class 'constraints'. */
   find_box(all_candidates ,0, con);
   con = best_box.get_constraints();
   return con;
   /* The return value shall contain all candidates, which
   intersect 'best_box'. */
}
void intersection_detection::find_box(box search_box, int
depth, constraints candidates)
```

```
{
   int axis;
   box left_half, right_half;
   constraints intersection();
   intersection.intersection_init(search_box, candidates); /
   * The object 'intersection' is initialized with all those
   candidates which intersect the 'search_box'. */
   constraints containing_set();
   containing_set.containing_init(search_box, candidates); /
   * The initialization of object 'containing_set' with all
   candidates which contain the 'search_box' */
   if (quality_factor(intersection) > q_best)
      if (intersection == containing_set)
      {
         q_best = quality_factor(intersection);
         best_box = search_box;
      }
      else
      {
         axis = depth % p; /* p is a private member of class inter-
         section_detection and holds the dimension of the trans-
         formation space. */
         left_half = search_box.left_half(axis);
         /* Assign the left half of 'search_box' along axis to
         'left_half' * /
         right_half = search_box.right_half(axis);
         /* Assign the right half of 'search_box' along axis to
         'right_half' */
         search_box(left_half, depth+1, intersection);
         search_box(right_half; depth+1, intersection);
      }
}
```

**Figure 37** Algorithm for the cluster search [93].

The algorithm is able to handle a transformation space of arbitrary dimension size $p$. The search is restricted to a window (search_box) $AW$ in every step. The algorithm starts initially with a window that encloses all uncertain areas (constraints or candidates). The algorithm uses the procedure 'quality_factor(intersection)' that calculates an entire quality factor from the elements belonging to an intersection that is transferred to the procedure. The estimated entire quality factor (upper boundary) of a window is used for the control of the algorithm. The upper boundary is determined for the actual window by the summation of those single quality factors whose uncertain areas cut the actual window (candidates). If the upper boundary is lower than a minimal value (q_min) or lower than the entire quality factor (q_best) found until now, the search will be terminated in an area determined by the actual window. If this is not true, the window is divided into a left and a right part and the search is performed recursive in these two parts. The recursive subdivision of the window happens sequentially along the single axes $i = 1, 2, \ldots, p$ of the transformation space. The abort criterion for the search is given by those contained uncertain areas in the window that cover the whole window. The best solution is then found. A second abort criterion stops the search if a determined recursion depth is gained. So a long runtime can be avoided when uncertain areas are close together, but do not touch. An actual implementation is now explained that extends the just-explained algorithm for the cluster search in some items:

1. The examination of the initial window happens automatically and is based on the minimal and maximal values of the transformation parameters from all single elements in the transformation space.
2. A recommendation value for the maximal recursion depth is derived. Calculation uses the size of the first window $AW_i$ and a minimal error value $\varepsilon_i$ for the single elements. The calculation is executed along the transformation axes $i$:

$$dr_{\max} = p \cdot \max_i\{h_i\} \quad \text{with} \quad h_i = \log_2\left(\frac{AW_i}{\varepsilon_i}\right),\ i = 1, 2, \ldots, p. \tag{5.6}$$

   The estimation presumes that the fragmentation of a window is only clever if the actual window size is not lower than the determined minimal error values of the single elements. If no valid solution exists at that time, then the actual solution can be considered as an approximation with a maximal error value in the magnitude of the single element's error values. $h_i$ denotes the number of necessary bisections of the transformation space per axis. This computed value is rounded up to an integer.
3. The original algorithm provides no estimation for the maximal runtime. Moreover, no restriction exists for the duration of the runtime. The determination of a maximal recursion depth is an instrument to restrict the runtime so that the use in real-time applications can be assured. This is further supported by the general limitation of the runtime. Therefore, suboptimal solutions are tolerated.
4. The uncertainty areas are modeled as squares. This is a further instrument to restrict the runtime, because intersections can be processed very simply. The transformation space is represented with a list of constraints. Every single element is described with an uncertainty area and a quality. The center coordinates and the measurements are attributes of the uncertainty areas. The measurements can be calculated with two diagonally opposite corners, because the uncertain areas are modeled as squares.
5. With several invocations of the procedure 'find_best_subset' and the masking of the found subsets, it is possible to generate several clusters in descending sequence with respect to the entire quality values. The subsets are stored in a data structure. New subsets are generated during the runtime of the algo-

rithm and old subsets are eliminated. The last window and the entire quality are stored. If a valid solution is represented by a subset, a flag is used to mark the subset.

6. Middle values are examined for valid subsets. For example, these can be the center coordinates of the last window. Further middle values are processed that are based on elements of that subset that represents a valid solution.

The selection of the uncertainty areas must be performed carefully, because the functionality of the cluster search depends strongly on the uncertainty areas. The position of the single elements in the transformation space $\mathbf{X}_T \subseteq R^p$ can be seen as a $p$-dimensional output variable $O \in \mathbf{X}_T$:

$$R^n \to R^p \quad \text{with} \quad op = \mathrm{Ft}(x). \tag{5.7}$$

In general, this is a nonlinear mapping that is processed with an inexact vectorial measurand $ip \in R^n$ and a covariance matrix $P'$. Let $y_k$ be a transformation hypothesis. The covariance matrix $P$ can be processed with an input vector $ip_k$:

$$P(ip_k) = J(ip_k) \cdot P' \cdot J^T(ip_k). \tag{5.8}$$

The Jacobi matrix is used to obtain linearity:

$$J(ip_k) = \left.\frac{\partial \mathrm{F}(x)}{\partial x}\right|_{ip=ip_k}. \tag{5.9}$$

The main diagonal of the matrix $P$ contains the variances of the transformation parameters. These variances determine the maximal extension of the uncertainty areas in the particular dimension of the transformation space. Dependencies exist between the transformation parameters. With these dependencies it is possible to get the uncertainty areas as $p$-dimensional polyhedrons by using regression lines, which are mostly a linear approximation of the real regression lines. The difficulty of this approach is the determination of the function $\mathrm{Ft}(x)$. This function is nonlinear and must be processed at runtime. Therefore, the uncertainty areas are modeled as squares in this implementation of the cluster search to save processing time. Now, a further error-propagation calculation is used that provides an estimation of the expected exactness of the middle transformation parameters of a cluster. The weighted averaging over $e$ single elements is written with the following formula:

$$O_k = (O_{k_i})^T_{i=1,2,\ldots,p}, \qquad k = 1, 2, \ldots, e. \tag{5.10}$$

$O_k$ with $p$ elements $O_{k_i}$ provides in combination with the single qualities $\kappa_k$ following output variable in the $p$-dimensional transformation space:

$$\bar{O} = (\bar{O}_i)^T_{i=1\ldots p} = \frac{\sum_{k=1}^{e} \kappa_k \cdot O_k}{\sum_{k=1}^{e} \kappa_k}. \tag{5.11}$$

The known standard deviations $\sigma_{O_{k_i}}$, $k = 1, 2, \ldots, e$, $i = 1, 2, \ldots, p$, of the single elements can be read from the covariance matrices and provide the mean error of the cluster elements' weighted mean value:

$$\sigma_{\bar{O}_i} = \frac{\sqrt{\sum_{k=1}^{e} \kappa_k^2 \cdot \sigma_{O_{k_i}}^2}}{\sum_{k=1}^{e} \kappa_k}. \tag{5.12}$$

### 5.6.3
### The Generation of a Pseudo-B-rep Representation from Sensor Data

It is now shown how a pseudo-B-rep representation can be generated from sensor data. Therefore, the three-dimensional contour will be segmented into plane partial contours. These will be the area primitives. The strategy consists of several steps. All possible binormals of the three-dimensional contour are calculated in the first step:

$$d_i = (x_C^{i-1} - x_C^{i}) \times (x_C^{i+1} - x_C^{i}). \tag{5.13}$$

The binormal is a vector that stands vertically on two successive polygon areas, see Figure 38.

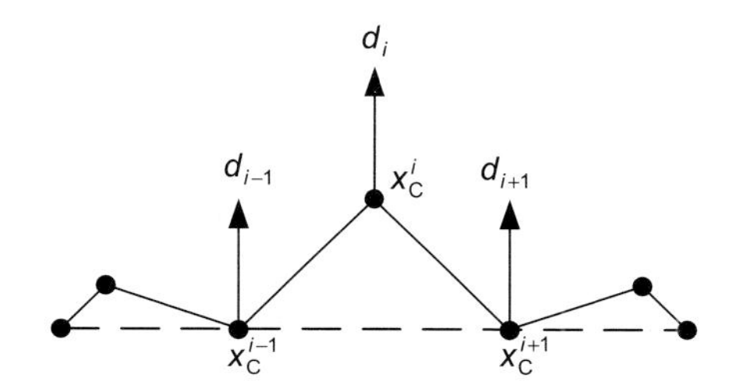


**Figure 38** The binormals of an object contour [93]

Because of disturbances it can be that polygon areas are segmented as independent contours. If several polygon areas meet in a corner, all involved areas must be considered for the calculation of the accompanying binormal. The coordinates of a binormal vector are also adjusted at an observer-centered camera coordinate system just like the corners. A cluster search procedure (see previous paragraph) is used to detect optimal plane partial polygons. Standardized binormals $d_i'$ are considered as elements of a three-dimensional transformation space

$$\mathbf{X}_{T_d'} \subset R^3: \tag{5.14}$$

$$d_i' = (x_{d'}^i, y_{d'}^i, z_{d'}^i)_{T_{d'}} = \frac{d_i}{\|d_i\|}. \tag{5.15}$$

The axes of $\mathbf{X}_{T_{d'}}$ are constituted by $X_{T_{d'}}$, $Y_{T_{d'}}$, and $Z_{T_{d'}}$. Additionally, a further vector is derived from every binormal vector. For these purposes the binormal vectors are rotated about 180° in the transformation space. This is necessary, because it is not possible to apply the terms 'inside' or 'outside' of an object before the object recognition has not taken place. So all area primitives are calculated twice. The selection of the correct area primitives can be gained in the process of object recognition. The object contours in the image are compared with the model contours. This happens from the right reading view and the side-inverted view. The side-inverted view can then be determined, because in that case the similarity to the model contour is rather poor. Normalized binormal vectors are weighted with the amount of a binormal vector:

$$\begin{aligned}\|d_i\| &= \left\|((x,y,z)_C^{i-1} - (x,y,z)_C^{i}) \times ((x,y,z)_C^{i+1} - (x,y,z)_C^{i})\right\| = \\ &\left\|((x,y,z)_C^{i-1} - (x,y,z)_C^{i}) \cdot ((x,y,z)_C^{i+1} - (x,y,z)_C^{i})\right\| \cdot \sin(\xi) = lg_i\end{aligned} \quad . \tag{5.16}$$

The weight factor includes the lengths of both polygon areas and angle $\xi$ that is generated from the two contours belonging to the polygon areas that meet in corners, where the binormal stands vertically. The weight assumes that an area can be processed more precisely in dependence of measurement errors of the corner coordinates the longer both primitive areas are and the closer the involved angle $\xi$ is to 90°. The whole quality of a cluster is determined by the sum of the single qualities. It is necessary to determine the dependency between the normalized binormal vectors and input parameters $ip$. The input parameters are necessary for the calculation of the normalized binormals. A cluster search is performed to obtain the size and form of uncertainty areas. The input parameters are derived from the camera coordinates of three neighboring corners of a polygon:

$$ip = ((x,y,z)_C^{i-1}, (x,y,z)_C^{i}, (x,y,z)_C^{i+1}). \tag{5.17}$$

A further vectorial transformation function $\mathrm{Tf}(x)$can be derived:

$$O = \frac{((x,y,z)_C^{i-1} - (x,y,z)_C^{i}) \times ((x,y,z)_C^{i+1} - (x,y,z)_C^{i})}{\left\|((x,y,z)_C^{i-1} - (x,y,z)_C^{i}) \times ((x,y,z)_C^{i+1} - (x,y,z)_C^{i})\right\|} = \mathrm{Tf}(x). \tag{5.18}$$

The covariance matrix $P$ is used in which the error measurements of the calculated corners are contained. The covariance matrix $P$ is the basis from which to process the covariance matrix $P'$ of the normalized binormals with an error-propagation calculation:

$$P'((x,y,z)_C^{i}) = J((x,y,z)_C^{i}) \cdot P \cdot J^{\mathrm{T}}((x,y,z)_C^{i}). \tag{5.19}$$

$J$ is the Jacobi matrix. The size and form of the uncertain areas can be examined with the covariance matrix $P'$. Because of the use of the error propagation, the normalized binormal vectors, which have lower single qualities, also have a larger unsure area. It is possible to obtain the corners of the discovered partial areas by using the binormal vectors at the end of the cluster search. With a simple plausibility test it can be ensured that only corners are aggregated to a polygon that are also associated in the original image data. Therefore, a threshold is used that determines the maximum allowable distance between corners. The distance should not be exceeded. It is now assumed that a plane partial polygon consists of $n$ corners. The mean value of a binormal vector can be computed by the addition of the particular cluster's binormals:

$$\hat{d} = \sum_{i=2}^{n-1} d_i . \tag{5.20}$$

The mean binormal $\hat{d}$ is a good approximation of a normal vector belonging to the plane primitive area. The precision of the examined binormal can be obtained from the covariance matrices in connection with the error-propagation calculation. Then the pseudo-B-rep representation can be constructed. The $m$ plane partial contours $BC_j, j = 1, 2, \ldots, m$, are now represented in a contour-centered coordinate system $\mathbf{X}_{BC_j}$:

$$CO_j = \left\{ (\overset{k_j}{x}, \overset{k_j}{y}, \overset{k_j}{z}, 1)_{BC_j} \middle| k = 1, 2, \ldots, n_j \right\} \quad \text{with} \quad \overset{k_j}{z}_{BC_j} = 0,\ j = 1, 2, \ldots, m. \tag{5.21}$$

The transformation matrix $H_{\mathrm{BC}_j,\mathrm{C}}$ shows the relations between the contour-centered coordinate system $\mathbf{X}_{\mathrm{BC}_j}$ and the camera-centered coordinate system $\mathbf{X}_\mathrm{C}$:

$$(\overset{k_j}{x}, \overset{k_j}{y}, \overset{k_j}{z})_\mathrm{C} = H_{\mathrm{BC}_j,\mathrm{C}} \cdot (\overset{k_j}{x}, \overset{k_j}{y}, \overset{k_j}{z})_{\mathrm{BC}_j} . \tag{5.22}$$

The position of the coordinate system $\mathbf{X}_{\mathrm{BC}_j}$ is determined arbitrarily, but the origin must be in the first corner $(\overset{1_j}{x}, \overset{1_j}{y}, \overset{1_j}{z})_\mathrm{C}$ of the plane partial polygon $B_j$ (I). The $Z_\mathrm{C}$-axis must be parallel to the direction of the mean binormal $\hat{d}^j_\mathrm{C}$ (II) and the $X_\mathrm{C}$-axis must be parallel to the $X_\mathrm{C} - Y_\mathrm{C}$ plane of the camera coordinate system (III). The transformation matrix $H_{\mathrm{BC}_j,\mathrm{C}}$ can be considered as the composition of column vectors in camera coordinates:

$$H_{\mathrm{BC}_j,\mathrm{C}} = (\overset{\mathrm{C}}{e_x}, \overset{\mathrm{C}}{e_y}, \overset{\mathrm{C}}{e_z}, \tau^\mathrm{C}). \tag{5.23}$$

Conditions I and II provide the following equations for the translation vector $\tau^\mathrm{C}$ and the direction vector $\overset{\mathrm{C}}{e_z}$ of the $Z_\mathrm{C}$-axis:

$$\tau_\mathrm{C} = \overset{1_j}{x_\mathrm{C}} \quad \text{and} \quad \overset{\mathrm{C}}{e_z} = \frac{\hat{d}^j_\mathrm{C}}{\left|\hat{d}^j_\mathrm{C}\right|} . \tag{5.24}$$

The direction vectors of the $X_\mathrm{C}$ and $Y_\mathrm{C}$ axes can be calculated by using condition III:

$$\overset{\mathrm{C}}{e_x} \cdot \overset{\mathrm{C}}{e_z} = \overset{\mathrm{C}}{e_x} \cdot \frac{\hat{d}^j_\mathrm{C}}{\left\|\hat{d}^j_\mathrm{C}\right\|} = 0 \quad \text{and} \quad \overset{\mathrm{C}}{e_y} = \overset{\mathrm{C}}{e_z} \times \overset{\mathrm{C}}{e_x} = \frac{\hat{d}^j_\mathrm{C}}{\left\|\hat{d}^j_\mathrm{C}\right\|} \times \overset{\mathrm{C}}{e_x} . \tag{5.25}$$

## 5.7
## Three-dimensional Reconstruction with Alternative Approaches

### 5.7.1
### Partial Depth Reconstruction

Only the relevant part of the observed scene is reconstructed in active vision to solve a specific task. An analogous strategy is followed for depth reconstruction. It is only that part of a reconstructed three-dimensional scene that is relevant to the specific problem. The strategy is called partial depth reconstruction. Depth reconstruction is necessary for object recognition, navigation, and manipulation. The depth information supports collision avoidance during the navigation and manipulation. Many objects have different appearances if they are viewed from different angles. In this case a reliable recognition is possible if the depth reconstruction of a three-dimensional object is available. If the system is equipped with knowledge about the scene, a lower level of reconstruction can be sufficient. The lower that knowledge is the more helpful, in general, can be the partial depth reconstruction that provides information for problem solving. It is also possible to use a special kind of illumination for the scene. Images that show the illuminated scene can then be analyzed. The light-section method uses a projector and a camera for depth reconstruction. The projector beams a line into the scene. The line is observed by the camera from a slightly displaced position. Because the positions of camera and projector are known, it is possible to calculate the depth information. A more precise depth reconstruction can be executed with a sophisticated approach that is called 'structured light'. The projector radiates vertical black and white light beams to an image sequence in this approach. The number of radiated light beams is doubled in successive images. The technique is more precise than the simpler light-section approach and requires a very precisely calibrated camera and projector. The simple approach is depth estimate from focusing that uses one rigid camera. If the focal length of the camera is variable, it is possible to take several images by differently adjusted focal lengths. It is possible to determine the distance for those pixels that are sharp in the respective image when the camera is calibrated suitably with regard to the relation between distance and focal length. The examination of the sharp pixels is difficult. The approaches that belong to the category 'relative depth' allow only the examination of relative relationships between objects in the depth. The approach 'occlusion analysis' ascertains only if an object is behind or in front of another object from the viewpoint of the observer. This can be discovered by using the statement that only an object that is nearer to the observer can occlude further remote objects. Some approaches provide only information about the form of an object, but no estimate for the depth like approach 'form reconstruction from shading or texture' [60].

Approach is now considered in more detail that belongs to the light-section methods.

### 5.7.2 Three-dimensional Reconstruction with Edge Gradients

The approach of Winkelbach and Wahl [95] is explained in this chapter. It needs a camera and two light stripe projections to reconstruct freeform. Two-dimensional gradient directions for reconstruction and generated surface normals are used. It is possible to apply this approach without calibration. The gradient directions are used to process the stripe angles in a two-dimensional stripe image as well as to obtain surface normals and ranges. To gain the local surface normal, it is necessary to take two images with rotated stripes with regard to both images. The reconstruction consists of two parts:

Part 1:
Two gray images are taken. Each image is illuminated with a different stripe projection with respect to the stripe orientation. A preprocessing is possibly necessary to eliminate disturbing information like texture or shading. A gradient operator is used to measure the local angles of stripe edges. Two-angle images are now available. But faulty angles and outliers remain.
Part 2:
The local surface slant (surface normal) is calculated with two stripe angles, which appear at the particular image pixel. The surface normals are used for the three-dimensional reconstruction of the object.

These steps in the reconstruction procedure are now considered in more detail. The preprocessing of a textured object is shown in Figure 39.

A textured object that was illuminated with stripes is shown in (a). The object is shown in (b) with ambient light. The image below left (c) shows the object with projector illumination. The difference between (a) and (b) can be seen in (d) and the difference between (b) and (c) in (e). Now the normalized stripe image can be found in (f) and in (g) the binary mask. The aim is the segmentation of the stripes in gray images. Therefore, gradient operators were used. Several operators like Sobel and Canny were evaluated to obtain an appropriate operator for the stripe-angle determination. This depends on the form of the object to be analyzed. Disturbed angles can be detected, because they appear mainly in homogeneous areas, where the gradient values are low. These angles are exchanged with interpolated data based on the neighborhood. Two strategies were used for the computation of the surface normals. The first strategy is a precise mathematical approach. The camera and the projectors are calibrated. The second strategy projects two stripe angles to one surface normal with a look-up table. The mathematical approach needs two angles $\Theta_1$ and $\Theta_2$ of the two rotated stripe projections and additionally the two-dimensional image coordinates $(x,y)_I^1$ and $(x,y)_I^2$. Also, the particular normals $n_1$ and $n_2$ are needed, see Figure 40.

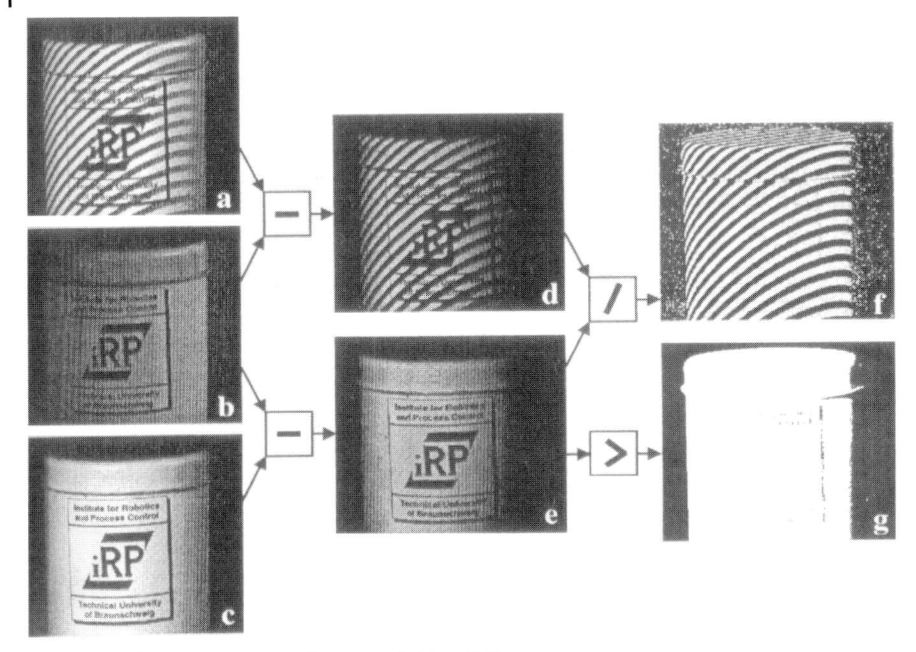


**Figure 39** The preprocessing of a textured object [95]

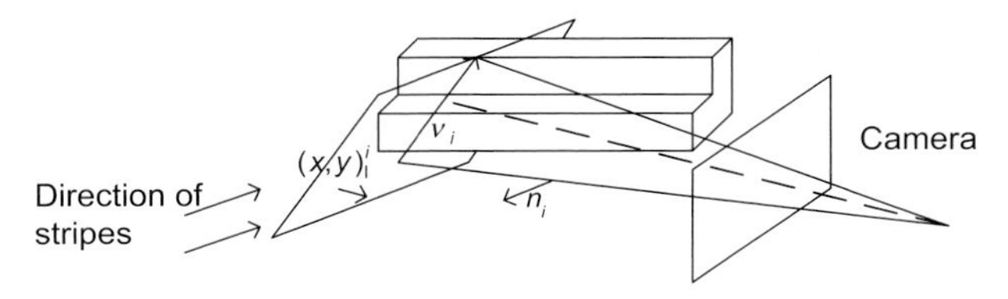


**Figure 40** Stripe projection [95]

Figure 40 also depicts the tangential direction vector $v_i$ of a stripe. The vector is orthogonal to $n_i$ and $(x,y)_I^i$. Then a formula is used to calculate the surface normal $\eta$:

$$\eta = (n_1 \times (x,y)_I^1) \times (n_2 \times (x,y)_I^2). \tag{5.26}$$

The creation of a look-up table is based on stereophotometry. First, the estimation of stripe angles is carried out. Therefore, two rotated stripe images are used as outlined before. Knowing the surface normals of a sphere is a prerequisite to filling the addresses of the look-up table with $\Theta_1$ and $\Theta_2$, whereby $\Theta_1$ and $\Theta_2$ are two angle values at each surface point of the sphere. It is possible to compute the missing values in the table with interpolation. A surface normal can be determined with a look-up table with respect to two stripe angles in the table.

### 5.7.3 Semantic Reconstruction

A knowledge-based approach for the three-dimensional reconstruction of landscapes in aerial images is discussed in this chapter according to [65]. Semantic reconstruction is used to improve the precision, reliability, and closeness to reality. This happens with an association between a three-dimensional model and sensor data. For these purposes two tasks must be executed:

1. The image is segmented into regions, which can be modeled apart.
2. An appropriate surface model is selected and model parameters are determined.

To obtain the segmentation of an image in regions with smooth surfaces, it is necessary to discover discontinuities. This approach uses an interpretation of the aerial image to derive knowledge about the model geometry. This will help to detect the discontinuities. The task of the surface model selection is supported by a priori knowledge to choose a suitable surface model. The model parameters are determined with the measured values. A geometric model can be taken from a generic model. The geometric model has to be chosen for a semantic object. The generic model supports the selection of an appropriate geometric model. The reconstruction of the three-dimensional geometry from the two-dimensional images is difficult because of occlusions, inaccurate measured values, and poor perspectives. The projection of a three-dimensional geometry into a two-dimensional image is simpler. So the model is projected into the image. The data are compared. The reconstruction procedure is iterative. The model parameters are modified until the projection fits best with the data. This approach is robust against initial errors in parameter estimation and against an inexact reconstruction procedure. Four steps are used for the reconstruction:

1. Model selection: A semantic scene description is used to get a suitable geometric surface model. The calculation of the parameters depends on the conditions for the model reconstruction.
2. Model-driven prediction: It is not often possible to match model and image directly. First, the model is decomposed. The parts are projected into the image.
3. Comparison: The transformation of the model into hypotheses for primitives arises with the prediction. These hypotheses can be compared with the segmented image primitives.
4. Data-driven modification: The results of the matching are transferred to the selected three-dimensional model. So a geometrical scene description is generated. The model parameters are constructed from the measured values of the particular primitive.

Steps one to three are repeated until the reconstruction of the three-dimensional object is completed, see Figure 41.

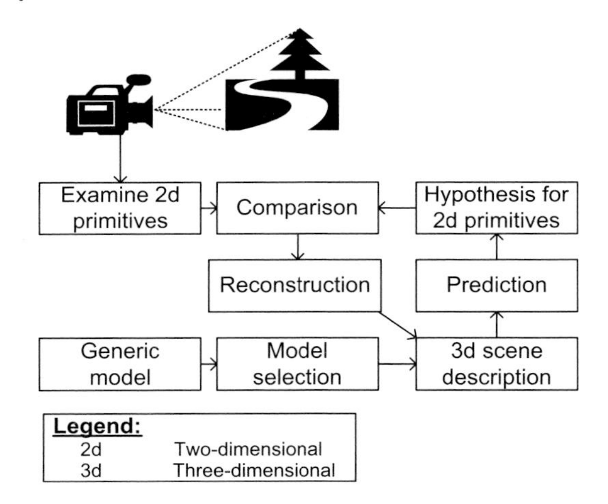


**Figure 41** Three-dimensional analysis with the projection of model data into image data [65]

The approach combines a model-driven and a data-driven procedure. The three-dimensional geometry of a semantic net represents the model that determines the aim of reconstruction. The surface description is stored in the nodes of the semantic net. An attribute calculation function is attached to the nodes. The function computes the surface data from the sensor data.

#### 5.7.3.1 **Semantic Reconstruction of Buildings**

Several algorithms for the reconstruction of buildings in aerial images have been developed [96–98]. We now consider in more detail an approach that executes a semantic reconstruction for buildings in aerial images. Therefore, a priori knowledge about the buildings is used for the reconstruction. A generic model of buildings is used, because various forms exist. So it should be possible to reconstruct different kinds of buildings, see Figure 42.

The semantic net in Figure 42 shows on the semantic level that buildings consist of several building wings. The 'is-a' relationship is used to represent the different roof shapes. The material level shows the material used for the bricks. For example, this can be clay. The geometry level holds the geometric forms of the building wings like a square stone. An initial model is used to reconstruct the square stones after the adaptation of the model parameters. The sensor level in the figure shows that it is necessary to detect a rectangle in the image that represents a square stone. Illuminated building roofs can possess different brightnesses because of different slopes. This can mean that building wings are represented by several partial rectangles, which must be merged. The rectangle represents an image region for the square stone with position parameters and form parameters. The fine adaptation between the square stone and the rectangle happens by matching between the rectangle's contour lines and the square stone edges. The extraction of a building uses the generic building model to verify the building hypothesis by performing a model-driven propagation to a partial rectangle. Therefore, the segmentation of rectangular regions is necessary. Figure 43 (a) shows the result of a texture analysis.

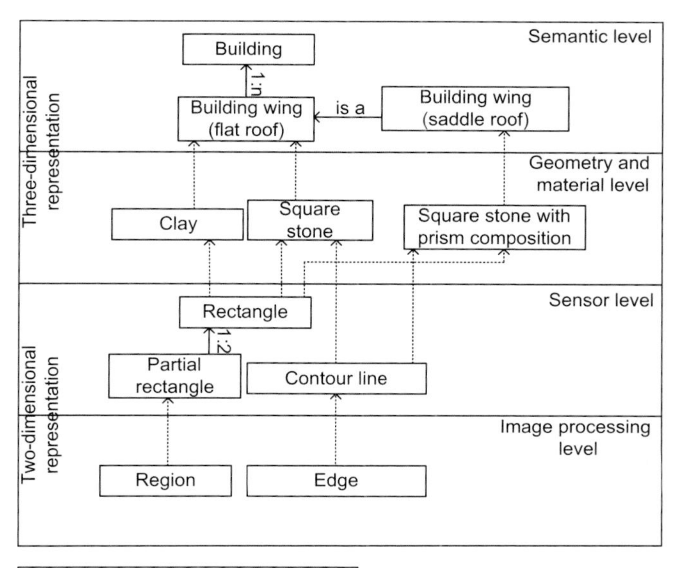


Figure 42 Semantic net for a building model [65]

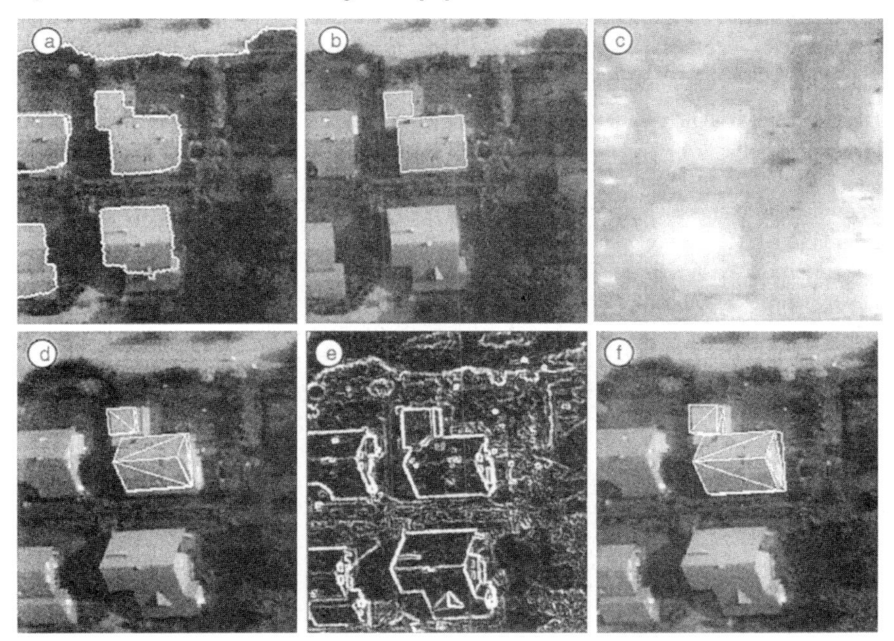


Figure 43 Matching between image data and model data [65]

The approximation of the detected regions with rectangles can be seen in (b). So it is possible to derive the initial length, width, and height of a square stone by using a depth map (c) and the size of the rectangle. Noise is visible in the depth map. This results in rather inaccurate parameters of the square stone (d). To obtain better results, the fine adaptation follows. The rectangle edges are compared with the square stone edges. The hypothesis for the square stone edges is projected into the gradient image (e). The comparison uses a similarity measure. This is the middle intensity of the amount-gradient image's pixels along the projected edge. Areas that are invisible are not recognized. If $A$ is the camera's line of sight, then the normal vector $n$ of the invisible areas is turned away from $A$:

$$A\,n > 0. \tag{5.27}$$

The adaptation of the parameter's orientation, length, width, and height of the square stone is then executed iteratively by adjusting the parameter's orientation, length, width, and height in the enumerated sequence. Now the determination of the building type follows. The hypothesis contains a building wing with saddleback roof. The search tree contains a scene description for a saddleback roof and a flat roof. The comparison of the image data with the square stone edges and the ridge results in the hypothesis saddleback roof, because the flat-roof model does not match to the image data. The building recognition is complete if all building wings are detected. This case is depicted in (f). The building consists of a building wing with saddleback roof and a further wing with a flat roof.

#### 5.7.3.2 Semantic Reconstruction of Streets and Forests

Three steps are applied for the reconstruction of streets and forests, which are illustrated in Figure 44. It starts with the segmentation and interpretation of an original image (a). The result will contain the streets and forests (b). A necessary depth map

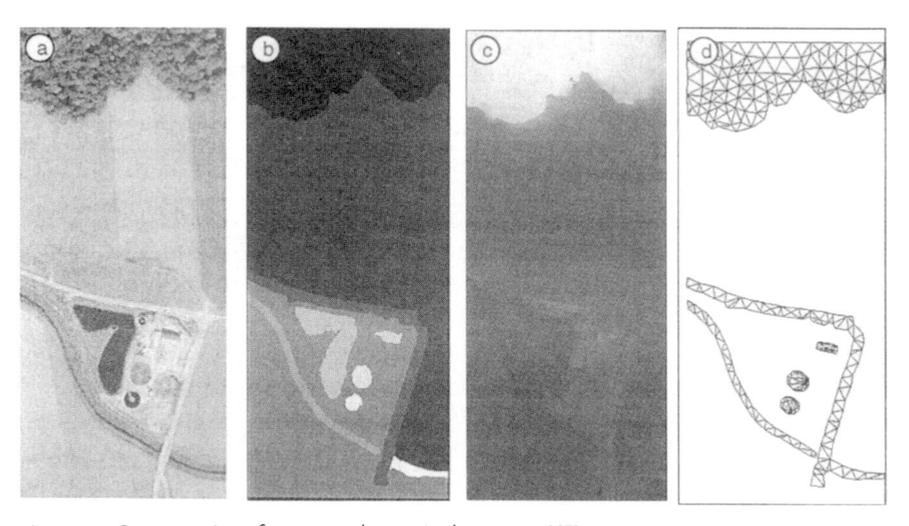


**Figure 44** Segmentation of streets and areas in three steps [65]

is generated in the second step (c). The depth map shows a thin plate for continuous regions. The last step represents the continuous regions detected in the depth map with a net consisting of triangles (d). These three steps are shown in the sequence.

The triangle net represents differences in elevation. Two steps are needed for the surface reconstruction:

1. The approximation of a continuous region contour with a closed polygon.
2. Every region is then filled up with triangles. The filling starts at the edge of the region.

Figure 45 depicts the two steps.

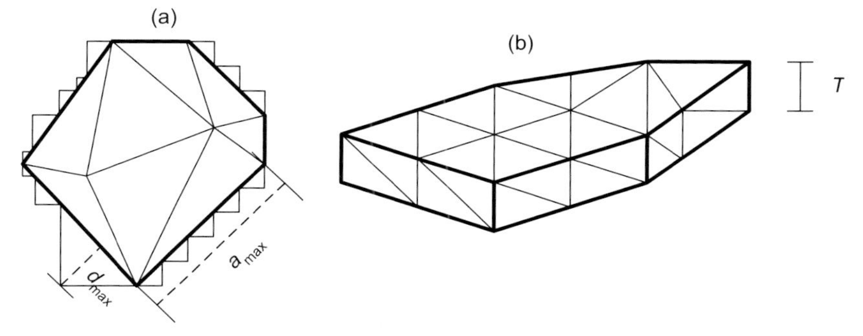


**Figure 45** The filling up of regions with triangles [65]

In (a) the closed polygon is shown that approximates the region contour. (b) demonstrates the filling of the region with triangles. The discontinuities in the depth map are reflected in the triangle net. The border contour runs between the pixels of neighboring regions, because every pixel belongs only to one region. The polygon approximation is a recursive procedure that generates longer border polygon vectors from shorter border contour vectors. The generation stops if the maximum allowable distance $d_{max}$ between border polygon and border contour is achieved or if the maximum length $a_{max}$ is exceeded. The parameter $a_{max}$ determines also the mesh size of the triangle net. The user has the possibility to select a threshold $T$. The threshold controls differences in elevation between neighboring regions. Such differences in elevation are considered in the model if the amount of difference, which is computed from two neighboring region heights measured at the $Z$-axis, exceeds the threshold. To obtain a scene description near to reality, semantics is used. For example, an edge of the forest can be finalized with a difference in elevation. This technique allows a qualitatively good reconstruction also when objects are occluded or some details are not reconstructed. For example, to get a realistic representation of an edge of the forest, the reconstructed landscape scene can be enriched with a grove from a graphic library.

#### 5.7.3.3 Texture

After the reconstruction of the geometrical surface is completed, determination of the photometry data is carried out. The aim is the modeling of the surface for the visualization. Therefore, the sensor image is projected as color texture onto the reconstructed surface. Matching between triangle and image cut is determined by a projection between two-dimensional image coordinates and the corresponding three-dimensional edge points of the triangle [65].

By using the ray theorem, it is possible to calculate that position in the image plane where the image plane is intersected from $A$. The ray theorem determines the length relations from the scalar product of the line of sight with the camera axes [65]:

$$p_{\mathrm{S}} = \frac{1}{(X-F)^{\mathrm{T}} \cdot z_{\mathrm{C}}} \cdot \begin{pmatrix} (X-F)^{\mathrm{T}} \cdot \dfrac{c}{g_x} x_{\mathrm{C}} \\ (X-F)^{\mathrm{T}} \cdot \dfrac{c}{g_y} y_{\mathrm{C}} \end{pmatrix} + h. \tag{5.28}$$

The parameters $p_{\mathrm{S}}$ and $h$ are measured in pixels, $g_x$ and $g_y$ in mm/pixel, and $X$ and $F$ in m. $x_{\mathrm{C}}$, $y_{\mathrm{C}}$, and $z_{\mathrm{C}}$ are without units.

The reconstruction of a scene point $X$ from a pixel $(x, y)_{\mathrm{S}}$ uses a stereo approach. The line of sight $A$ belonging to the pixel $(x, y)_{\mathrm{S}}$ is defined as the unit vector [65]:

$$A_{(x,y)_{\mathrm{S}}} = \frac{c \cdot z_{\mathrm{C}} + g_x \cdot x_{\mathrm{S}} \cdot x_{\mathrm{C}} + g_y \cdot y_{\mathrm{S}} \cdot y_{\mathrm{C}}}{\left| c \cdot z_{\mathrm{C}} + g_x \cdot x_{\mathrm{S}} \cdot x_{\mathrm{C}} + g_y \cdot y_{\mathrm{S}} \cdot y_{\mathrm{C}} \right|}. \tag{5.29}$$

It is necessary to know the distance $d$ between the scene point $X$ and the projection center $F$. Stereotriangulation is executed with two aerial images to determine $d$ [65]. The two aerial images, which are necessary for the triangulation, have an overlap of 60 %. The position of the corresponding pixels in the two images can be calculated if the orientation of the camera is known at the recording time. The orientation data $(F, z_{\mathrm{C}}, H, L)_{\mathrm{L}}$, $(F, z_{\mathrm{C}}, H, L)_{\mathrm{R}}$ of the two recordings are examined with registration. So the scene point $X$ is processed from the intersection of the lines of sight $A_{\mathrm{L}}$ and $A_{\mathrm{R}}$ belonging to pixels $p_{\mathrm{L}}$ and $p_{\mathrm{R}}$. The camera registration is inaccurate. This means that in many cases the two lines of sight will not intersect. Therefore, a minimal distance $|d'|$ between both warped lines is required [65]:

$$X = F_{\mathrm{L}} + d_{\mathrm{L}} \cdot A_{\mathrm{L}} = F_{\mathrm{R}} + d_{\mathrm{R}} \cdot A_{\mathrm{R}} + d' \tag{5.30}$$

with an additional condition: $A_{\mathrm{L}}^{\mathrm{T}} \cdot d' = A_{\mathrm{R}}^{\mathrm{T}} \cdot d' = 0.$ (5.31)

The multiplication of this formula by $A_{\mathrm{L}}^{\mathrm{T}}$ and $A_{\mathrm{R}}^{\mathrm{T}}$, respectively, provides two scalar equations that can be resolved to $d_{\mathrm{L}}$ [65]:

$$d_{\mathrm{L}} = \frac{(F_{\mathrm{L}} - F_{\mathrm{R}})^{\mathrm{T}} \cdot (A_{\mathrm{L}}^{\mathrm{T}} \cdot A_{\mathrm{R}}^{2} - A_{\mathrm{L}})}{1 - (A_{\mathrm{L}}^{\mathrm{T}} \cdot A_{\mathrm{R}})}. \tag{5.32}$$

### 5.7.4 Mark-based Procedure

Mark-based procedures enable processing of the object position without using information about the camera's position and the alterations in images, which result from the alteration of the camera position. The examination of the object position and the orientation occurs by using knowledge about the object, which must be known additionally for the computation if the camera position is not known. Mark-based procedures use features of rotation bodies like spheres and cylinders. They also use line features, for example, of circular discs. Point features are also used. If point features are used, three points at least are necessary to explore the position and orientation of an object. These three points form a triangle. If the form and the size of the triangle are known, the object's position and orientation can be computed. Lang gives an introduction into mark-based procedures and explains some methods. The elucidations are summarized in the next sections [33].

#### 5.7.4.1 Four-point Procedure

The four-point procedure is a mark-based procedure that uses four points $X_k$, $k = 1, \ldots, 4$, for the determination of the object position. Distance $d_{ij}$, $i, j = 1, \ldots, 4$, between two points $X_i$ and $X_j$, which are mounted on the object, is known for every point pair. So it is possible to compute the object position. Therefore, the distance between the points that are projected onto the camera sensor is taken. The approach consists of two steps. In the first step the position of the four points is determined in relation to the camera coordinates. In the second step a transformation matrix between the object coordinate system and the camera coordinate system is calculated, see Figure 46.

The calculation of the camera coordinates uses volume computation of tetrahedra. A tetrahedron is a pyramid with a triangular base. The computation of the tetrahedra takes place by connecting the four points $X_1$, $X_2$, $X_3$, and $X_4$ of the real object and the origin of the camera coordinate system $\mathbf{X}_C$. The base of the rectangle is divided into four triangles to get four tetrahedra with respect to the camera coordinate origin. Similarly it is possible to determine four tetrahedra between the origin of the camera coordinate system and the camera sensor. The object points' camera coordinates can now be determined with the edge lengths of the tetrahedra.

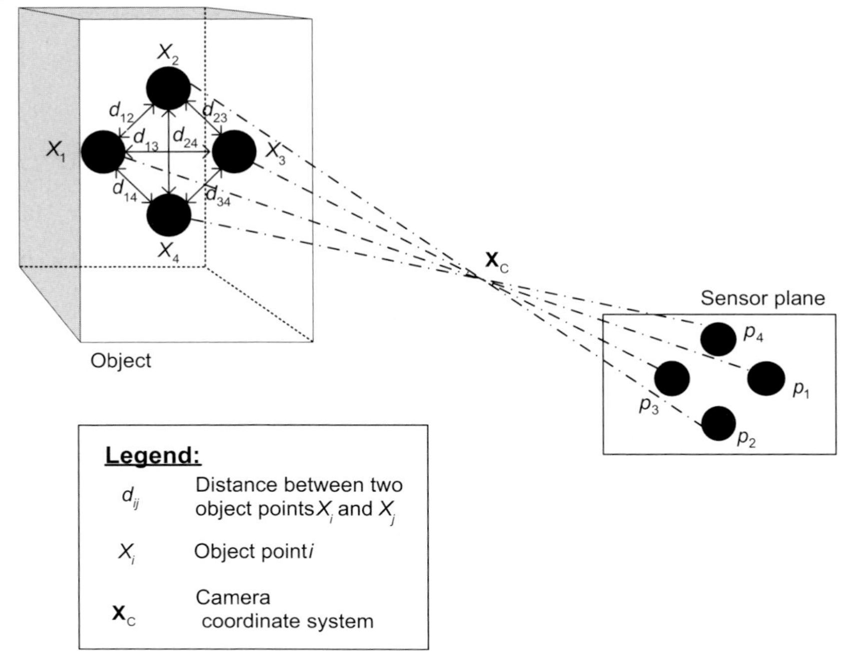


**Figure 46** Building of a tetrahedron [33]

5.7.4.2 **Distance Estimation**

Simple approaches for distance estimation between the camera and the object exist. A surface feature is necessary for the distance estimation. Here, this is the object size. The distance estimation uses the fact that an object that is nearer to the camera appears larger in the image than an object that is more remote. Of course, this is only a heuristic method, because the object size in the image depends also on the angle from which the image is taken. If the camera takes an image from a more skewed perspective with regard to the object, the computed values are rather inaccurate. To determine the object size in the image, it is possible to use an approach based on mark points that are attached to the object. Distances between the mark points in the image are used to calculate the object size. For example, if four mark points $p_i$, $i = 1, 2, 3, 4$, of different sizes are used, all these four points should be visible in the image to get a good approximation of the real object size. The actual object size $\vartheta$ can be calculated with the following formula if four mark points are used:

$$\vartheta = \frac{d_{12}+d_{34}}{2} \cdot \frac{d_{24}+d_{31}}{2}. \tag{5.33}$$

$d_{12}$, $d_{24}$, $d_{34}$, and $d_{31}$ are distances between the mark points. For example, $d_{12}$ represents the distance between mark points $p_1$ and $p_2$. The approach requires that the four mark points represent the corners of a rectangle. Point $p_1$ should represent the upper-left corner, $p_2$ the upper-right corner, $p_3$ the lower-left corner, and $p_4$ the lower-right corner. The rectangle's border has to be generated by the distances. The mark points' plane must be orthogonal to the principal axis. If these restrictions are not met, approximate values for the object size only can be computed. It is also necessary that the used mark points differ in size. The computation of the actual object size using camera coordinates proceeds as follows:

$$\vartheta = \frac{4 \cdot x_C \cdot y_C}{(z_C)^2} \cdot (b)^2. \tag{5.34}$$

The estimation of the object distance between the camera and the object is performed by comparison between the actual object size $\vartheta$ and the object size $\vartheta'$. The object size $\vartheta'$ is trained in a teaching process and can be processed with the following formula:

$$\vartheta' = \frac{4 \cdot x_C \cdot y_C}{(z'_C)^2} \cdot (b')^2. \tag{5.35}$$

The focal length of the camera during the teaching phase is denoted by $b'$. $z'_C$ is the distance of the object during the teaching process. To determine the distance between the actual object and the camera, the following formula must be used:

$$z_C = \frac{b}{b'} \cdot \sqrt{\frac{\vartheta'}{\vartheta}} \cdot z'_C \tag{5.36}$$

To obtain the distance $z'_C$ of the teaching process, measurements can be executed in advance. $\vartheta'$ can be computed by using the mark-point coordinates in the teaching process. A further approach uses the real object size $\vartheta^{\text{real}}$ to calculate the distance between the camera and the object. Mark points can also be used here. The actual object size can be determined with the following formula:

$$\vartheta = \frac{\vartheta^{\text{real}}}{(z_C)^2} \cdot b^2. \tag{5.37}$$

The resolution of the formula to $z_C$ yields the distance:

$$z_C = \sqrt{\frac{\vartheta^{\text{real}}}{\vartheta}} \cdot b. \tag{5.38}$$

# 6
# Stereo Vision

Stereo vision is based on the human visual apparatus that uses two eyes to gain depth information. Stereo vision produces a doubling of the processing time in comparison to a monocular visual apparatus, because two images must be analyzed. Therefore, it is recommended to implement stereo approaches with parallel algorithms. For example, parallelism can be realized with threads. Threads are objects that can be executed in parallel. Therefore, every camera can be represented in a software program by one object running as a thread. If such a software program uses a single processor machine, quasiparallelism is realized, because the threads have to share the single processor. This can happen by the alternating use of the processor, but the realized execution sequence of the threads depends strongly on the threads' priority. The programmer can determine the priority as a function of the used thread library. Real parallelism can be realized if a multiprocessor machine is used. For example, if two threads are running on a double-processor machine, whereby every thread represents a camera, it is possible that both threads use a different processor during the program execution.

The textbook of Sonka, Hlavac, and Boyle gives an introduction to stereo vision with its geometrical relationships, which result from the particular configuration of the cameras. It explains the projection of points from three-dimensional space into two-dimensional space. Applications with moving cameras are considered. The three-dimensional reconstruction with two or more cameras of a scene point from two-dimensional image data is illustrated. The book also discusses the finding of corresponding pixels in different images. These subjects are summarized in the next sections [63].

## 6.1
## Stereo Geometry

To obtain the depth information by the use of at least two cameras, it is necessary to have some data about the geometry of the cameras used, see Figure 47.

*Robot Vision: Video-based Indoor Exploration with Autonomous and Mobile Robots.* Stefan Florczyk

ISBN: 3-527-40544-5

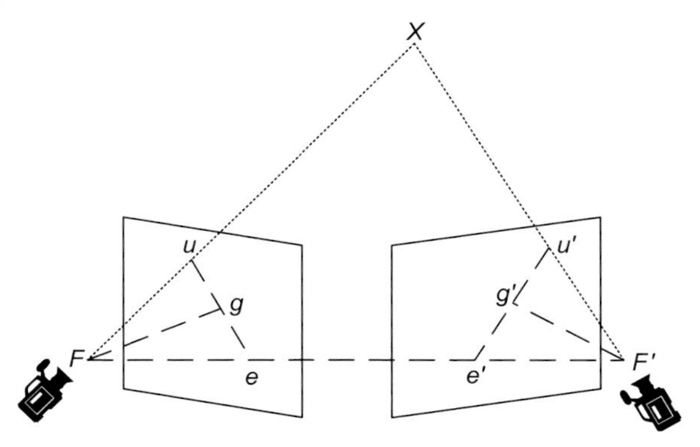


**Figure 47** Geometry in stereo vision [63]

The first step in stereo vision is the calibration of the cameras to determine each camera's line of sight. To get depth information in stereo vision, it is required that two lines of sight intersect in the scene point $X$ for which the depth information is to be processed. The last step in stereo vision is the examination of three-dimensional coordinates belonging to the observed scene point. Figure 47 shows the geometry of a stereo-vision system. The two optical centers $F$ and $F'$ are associated with a baseline. The lines of sight belonging to $F$ and $F'$ intersect at the point $X$ and generate a triangular plane that intersects every image plane in epipolar lines $g$ and $g'$. The projections $u$ and $u'$, respectively, of the scene point $X$ can be found on these two lines. All possible positions of $X$ lie on the line $FX$ for the left image and on the line $F'X$ for the right image. This is reflected in the left camera's image plane for $F'X$ by the line $g$ and in the right camera's image plane for the line $F'X$ by the line $g'$. The epipoles $e$ and $e'$ also lie on the lines $g$ and $g'$ that are intersected by the baseline. These geometric relations can be used to reduce the search dimension from two dimensions to one dimension.

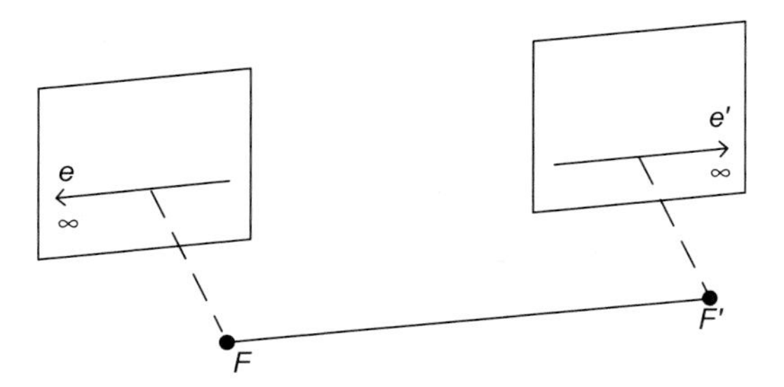


**Figure 48** Canonical stereo configuration [99]

Figure 48 shows the simple stereo configuration that is called canonical. The baseline matches to the horizontal coordinate axis. The two cameras' lines of sight are parallel. The configuration has the consequence that the two epipoles $e$ and $e'$

are infinite. The epipolar lines run horizontally. This configuration is used in a manual method operated by a human. The human analyses the horizontal lines in the image to find matching scene points. The conclusion for the development of an automated procedure is that matching points can be found by the examination of horizontal lines rather than by analyzing arbitrary lines. With image rectification a general stereo approach with nonparallel epipolar lines can be turned into a canonical stereo configuration by using appropriate formulas. The use of formulas for rectification requires resampling that decreases the resolution of images. Therefore, if a vision-based application needs high-resolution images, a general stereo approach should be used. Figure 49 shows the canonical stereo configuration.

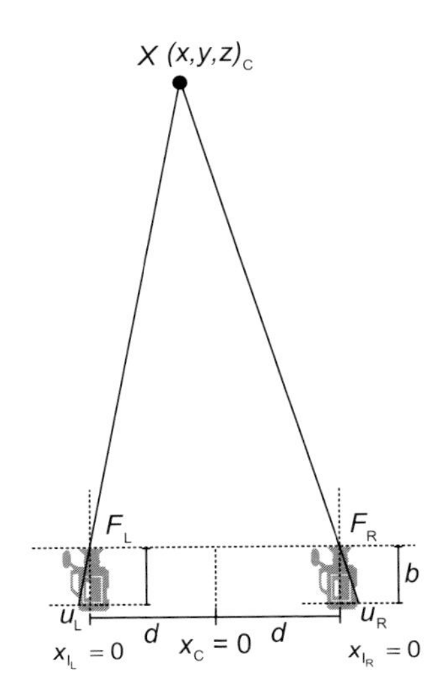


**Figure 49** Stereo geometry in canonical configuration [63]

It can be seen in the figure that the principal axes of the two cameras are parallel. The two cameras have the distance of $2d$ between them. $u_L$ and $u_R$ are the projections of the point $X$ onto the particular camera's image plane. $z_C$ represents the distance between the camera and the scene point. The horizontal distance is represented by $x_C$. $y_C$ is not illustrated in Figure 49. Coordinate systems $\mathbf{X}_{I_L}$ and $\mathbf{X}_{I_R}$ exist for every image plane. $\mathbf{X}_{I_L}$ represents the image plane of the camera on the left-hand side and $\mathbf{X}_{I_R}$ the image plane on the right-hand side. There is an inequality between $\mathbf{X}_{I_L}$ and $\mathbf{X}_{I_R}$ resulting from different camera positions. The value for $z_C$ can be determined with base geometric processing. $u_L$, $F_L$ and $F_L$, $X$ are hypotenuses of orthogonal triangles that look very similar. $d$ and $b$ are positive numbers, $z_C$ is a positive coordinate. $x_C$, $u_L$, and $u_R$ are coordinates with positive or negative values. These considerations result in the following formula for the left-hand side:

$$\frac{u_L}{b} = -\frac{d+x_C}{z_C}. \qquad (6.1)$$

The formula for the right-hand side is very similar:

$$\frac{u_R}{b} = \frac{d - x_C}{z_C}. \tag{6.2}$$

Now $x_C$ will be eliminated from the equations:

$$z_C(u_R - u_L) = 2db. \tag{6.3}$$

and resolved to $z_C$:

$$z_C = \frac{2db}{u_R - u_L}. \tag{6.4}$$

## 6.2 The Projection of the Scene Point

Until now it was necessary to have a camera configuration with two parallel principal axes. This restriction will be given up in next approach. The translation vector $\tau$ is used to transform between the left camera center $F$ and the right camera center $F'$. This affects a mapping between the coordinate system belonging to the camera on the left-hand side and the other coordinate system on the right-hand side. Additionally, the rotation matrix $T$ is necessary to calculate the projection. The center $F$ is the origin in the left coordinate system. The calibration matrices $\Psi$ and $\Psi'$ of the left and right cameras are needed to apply the following formula:

$$\tilde{u} = \begin{bmatrix} x_A \\ y_A \\ z_A \end{bmatrix} = [\Psi T | - \Psi T \tau] \begin{bmatrix} X \\ 1 \end{bmatrix} = D \begin{bmatrix} X \\ 1 \end{bmatrix} = D\tilde{X}. \tag{6.5}$$

The calibration matrix describes the technical parameters of the camera. The calibration matrix does not collect external parameters, which depend on the orientation of the camera. The external parameters are registered with the rotation matrix $T$ and translation vector $\tau$. The determination of the camera calibration matrix is described in Chapter 7.

The scene point $X$ is expressed in homogeneous coordinates $\tilde{X} = [X, 1]$ in this formula. This holds also for the projected point $\tilde{u} = [x_A, y_A, z_A]^T$ that has the two-dimensional Euclidean equivalent $u = [x_A, y_A]^T = [x_A/z_A,\ y_A/z_A]^T$. So it is possible to write $\tilde{u}$ by using a $3 \times 4$ matrix. A $3 \times 3$ submatrix on the left-hand side of the $3 \times 4$ matrix describes a rotation. The remaining column on the right-hand side represents a translation. A vertical line denotes this. $D$ is the projection matrix, also known as the camera matrix. If the projection of the camera from three-dimensional space into a two-dimensional plane is expressed in a simple manner, the following projection matrix may be sufficient:

$$D = \begin{bmatrix} 1 & 0 & 0 & 0 \\ 0 & 1 & 0 & 0 \\ 0 & 0 & 1 & 0 \end{bmatrix}. \tag{6.6}$$

With this projection matrix only a very abstract model can be expressed without consideration of any camera parameters of the used camera. This simple matrix belongs to the normalized camera coordinate system. As noted before, Equation (6.5) can now be used to derive the left and right projections $u$ and $u'$ of the scene point $X$:

$$u \cong [\Psi|0]\begin{bmatrix} X \\ 1 \end{bmatrix} = \Psi X, \tag{6.7}$$

$$u' \cong [\Psi' T| - \Psi' T\tau]\begin{bmatrix} X \\ 1 \end{bmatrix} = \Psi'(TX - T\tau) = \Psi' X'. \tag{6.8}$$

The symbol $\cong$ in the formula represents a projection up to an unknown scale. In the following the subscripts R and L are used to denote the difference between left camera's view and the right camera's view. The symbol $\times$ denotes the vector product. Now the coordinate rotation for right and left scene points is be given by:

$$X'_{\mathrm{R}} = T\,X'_{\mathrm{L}} \quad \text{and} \quad X'_{\mathrm{L}} = T^{-1}\,X'_{\mathrm{R}}. \tag{6.9}$$

It is valid that the vectors $X$, $X'$, and $\tau$ are coplanar. This is described with the following formula:

$$X_{\mathrm{L}}^{\mathrm{T}}(\tau \times X'_{\mathrm{L}}) = 0. \tag{6.10}$$

If equations $X_{\mathrm{L}} = \Psi^{-1}u$, $X'_{\mathrm{R}} = (\Psi')^{-1}u'$, and $X'_{\mathrm{L}} = T^{-1}(\Psi')^{-1}u'$ are substituted, the next formula results:

$$(\Psi^{-1}u)^{\mathrm{T}}(\tau \times T^{-1}(\Psi')^{-1}u') = 0. \tag{6.11}$$

Now the vector product is replaced by matrix multiplication. Let $\tau = [\tau_x, \tau_y, \tau_z]^{\mathrm{T}}$ be the translation vector. A skew symmetric matrix $W(\tau)$ is generated with the translation vector [100]:

$$W(\tau) = \begin{bmatrix} 0 & -\tau_z & \tau_y \\ \tau_z & 0 & -\tau_x \\ -\tau_y & \tau_x & 0 \end{bmatrix}. \tag{6.12}$$

Whereby a matrix $W$ is skew symmetric if $W^{\mathrm{T}} = -W$. $\tau \neq 0$ must be fulfilled. Rank $(W)$ symbolizes the number of linearly independent lines in $W$. Rank $(W) = 2$ holds if and only if $\tau \neq 0$. The next equation shows the displacement of a vector product by the multiplication of two matrices. $M$ is any regular matrix:

$$\tau \times M = W(\tau)\,M. \tag{6.13}$$

Knowing this, the formula (6.11) can be rewritten in the following manner:

$$(\Psi^{-1}u)^{\mathrm{T}}(W(\tau)T^{-1}(\Psi')^{-1}u') = 0. \tag{6.14}$$

Now, additional rearrangements are applied:

$$u^{\mathrm{T}} (\Psi^{-1})^{\mathrm{T}} W(\tau) T^{-1} (\Psi')^{-1} u' = 0. \tag{6.15}$$

The formula results in the fundamental matrix $B$ that represents the middle part of the previous formula:

$$B = (\Psi^{-1})^{\mathrm{T}} W(\tau) T^{-1} (\Psi')^{-1}. \tag{6.16}$$

In the end $B$ is inserted into the formula (6.15):

$$u^{\mathrm{T}} B\, u' = 0. \tag{6.17}$$

This is the bilinear relation also known as the Longuet–Higgins equation.

## 6.3 The Relative Motion of the Camera

We now consider moving cameras in space by known calibration matrices $\Psi$ and $\Psi'$ that are used for the normalization of measurements in both images. Let $\breve{u}$ and $\breve{u}'$ be the normalized measurements:

$$\breve{u} = \Psi^{-1} u, \quad \breve{u}' = (\Psi')^{-1} u'. \tag{6.18}$$

These formulas can be inserted into (6.15):

$$\breve{u}^{\mathrm{T}} W(\tau) T^{-1} \breve{u}' = 0. \tag{6.19}$$

Now a term in the formula is substituted by $E = W(\tau)T^{-1}$. $E$ is the essential matrix:

$$\breve{u}^{\mathrm{T}} E\, \breve{u}' = 0. \tag{6.20}$$

Also a bilinear relation originates in the case of stereo vision with nonparallel axes. $E$ contains the information about the calibrated camera's relative motion from a first to a second point. $E$ can be estimated using image measurements. The properties of $E$ are now listed:

1. $E$ has rank 2.
2. $\tau' = T\tau$. Then $E\tau' = 0$ and $\tau^{\mathrm{T}} E = 0$, whereby $\tau$ is the translation vector.
3. It is possible to decompose $E$ into $E = M\, ZN^{\mathrm{T}}$ for a diagonal matrix $Z$ with SVD (singular-value decomposition):

$$Z = \begin{bmatrix} c & 0 & 0 \\ 0 & c & 0 \\ 0 & 0 & 0 \end{bmatrix}. \tag{6.21}$$

$Z = diag[c, c, 0]$, whereby $diag[c_1, c_2, \ldots]$ is a diagonal matrix with diagonal $c_1, c_2, \ldots$.

It is now presumed that the matrix $E$ has been guessed, and rotation $T$ and translation $\tau$ between the two views is examined. As shown before, the essential matrix $E$ is the product of $W(\tau)$ and $T^{-1}$. The next formula shows therefore Equation (6.19) for the left and right views:

$$\breve{u}^{\mathrm{T}} W(\tau) T^{-1} \breve{u}' = 0, \quad \breve{u}'^{\mathrm{T}} T W(\tau) \breve{u} = 0. \tag{6.22}$$

The equation $E = TW(\tau)$ is also valid. To apply SVD for decomposition, the next two matrices are necessary [100]:

$$M = \begin{bmatrix} 0 & 1 & 0 \\ -1 & 0 & 0 \\ 0 & 0 & 1 \end{bmatrix}, \quad Z = \begin{bmatrix} 0 & -1 & 0 \\ 1 & 0 & 0 \\ 0 & 0 & 0 \end{bmatrix}. \tag{6.23}$$

Using the matrix $M$ for SVD provides the following equations for the rotation matrix $T$:

$$T = IMN^{\mathrm{T}} \quad \text{or} \quad T = IM^{\mathrm{T}}N^{\mathrm{T}}. \tag{6.24}$$

SVD can also be used to obtain the components of the translation vector from the matrix $W(\tau)$ by using the matrix $Z$ [100]:

$$W(\tau) = NZN^{\mathrm{T}}. \tag{6.25}$$

## 6.4 The Estimation of the Fundamental Matrix *B*

The fundamental matrix has some properties that will be mentioned here:

- As noted before, the essential matrix $E$ has rank two. Inserting $E$ into Equation (6.16) provides $B = (\Psi^{-1})^{\mathrm{T}} E (\Psi')^{-1}$. The fundamental matrix $B$ then also has rank two if the calibration matrices are regular.

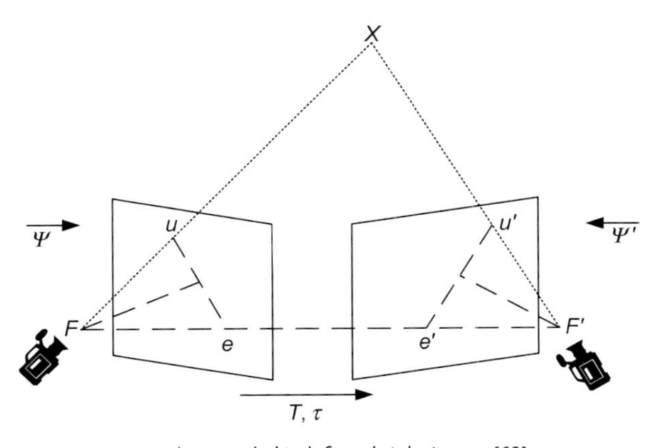


**Figure 50** Epipoles $e$ and $e'$ in left and right image [63]

- Figure 50 shows a left and a right image with the particular epipoles $e$ and $e'$. For these epipoles the following holds:

$$e^{\mathrm{T}} B = 0 \quad \text{and} \quad B\, e' = 0. \tag{6.26}$$

- SVD can also be applied to the fundamental matrix $B = M\, ZN^{\mathrm{T}}$:

$$Z = \begin{bmatrix} c_1 & 0 & 0 \\ 0 & c_2 & 0 \\ 0 & 0 & 0 \end{bmatrix}, \quad c_1 \neq c_2 \neq 0. \tag{6.27}$$

Seven corresponding points in the left and the right image are sufficient for estimation of the fundamental matrix $B$ with a nonlinear algorithm. But this algorithm is numerically unstable. A long runtime can be needed for the computation of a result [101]. An eight-point algorithm needs at least eight noncoplanar corresponding points and works with a linear algorithm that warrants a suitable processing time. Using more than eight points has the advantage that the processing is robust against noise. To guarantee an appropriate runtime it is recommended to perform normalizations of values, for example, for translation and scaling. The equation for the fundamental matrix $B$ is now written with subscript A that refers to the image affine coordinate system as introduced before:

$$u_{\mathrm{A}_i}^{\mathrm{T}} \quad u'_{\mathrm{A}_i} = 0, \quad i = 1, 2, \ldots, 8. \tag{6.28}$$

An image vector in homogeneous coordinates can be written in the manner $u^{\mathrm{T}} = [x_{\mathrm{A}_i}, y_{\mathrm{A}_i}, 1]$. Eight unknowns must be determined in the $3 \times 3$ fundamental matrix $B$. Eight matrix equations can be constructed by using the eight known correspondence points:

$$[x_{\mathrm{A}_i}, y_{\mathrm{A}_i}, 1]\; B \begin{bmatrix} x'_{\mathrm{A}_i} \\ y'_{\mathrm{A}_i} \\ 1 \end{bmatrix} = 0. \tag{6.29}$$

Now rewriting the fundamental matrix's elements as a column vector with nine elements creates linear equations $v^{\mathrm{T}} = [w_{11}, w_{12}, \ldots, w_{33}]$:

$$\begin{matrix} x_{\mathrm{A}_i} x'_{\mathrm{A}_i} & x_{\mathrm{A}_i} y'_{\mathrm{A}_i} & x_{\mathrm{A}_i} & y_{\mathrm{A}_i} x'_{\mathrm{A}_i} & y_{\mathrm{A}_i} y'_{\mathrm{A}_i} & y_{\mathrm{A}_i} & x'_{\mathrm{A}_i} & y'_{\mathrm{A}_i} & 1 \\ & & & & \vdots & & & & \end{matrix} \begin{bmatrix} w_{11} \\ w_{12} \\ \vdots \\ w_{33} \end{bmatrix} = 0. \tag{6.30}$$

$M$ symbolizes the matrix on the left-hand side:

$$Mv = 0. \tag{6.31}$$

Assuming that a favored case with no noise is to be handled, $M$ has rank eight. Using data from real measurements we can construct an overdetermined system of linear equations that can be solved with the least-squares method. But the estimation of the fundamental matrix $B$ can be erroneous if mismatches exist between corresponding points.

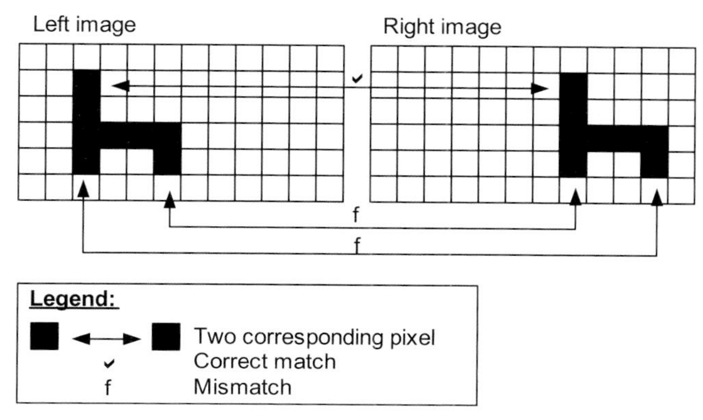


**Figure 51** Mismatches between corresponding points [63]

Figure 51 shows two images with pixel grids. The contour of a chair can be viewed in both images. Two pixels in both images, which are part of a backrest, have been matched correctly in contrast to four pixels occurring in both chair legs.

The mismatch of pixels requires the elimination of the faulty matches. This can be executed by using the least median of squares:

$$\min_{v}(v^{\mathrm{T}} M^{\mathrm{T}} M v) \rightarrow \min_{v}[\mathrm{median}(\|Mv\|^{2})]. \tag{6.32}$$

As mentioned before, the fundamental matrix $B$ is of rank two. Equation (6.31) does not always provide a matrix of rank two. Therefore, $B$ is changed to $\hat{B}$. The matrix $\hat{B}$ minimizes the Frobenius norm $\|B - \hat{B}\|$. To process $\hat{B}$, SVD can be applied to decompose $B = MZN^{\mathrm{T}}$, $Z = \mathrm{diag}[c_1, c_2, c_3]$, $c_1 \geq c_2 \geq c_3$. Now $\hat{B} = M\,\mathrm{diag}[c_1, c_2, 0]N^{\mathrm{T}}$ is obtained.

## 6.5 Image Rectification

Epipolar lines can be used to find corresponding points in stereo vision. Often epipolar lines will not be parallel in both images, because the principal axes of the cameras used are not also parallel. It is possible to apply an adjustment to obtain parallel optical lines. This strategy has the advantage that parallel epipolar lines support the search for corresponding points in a better way. The correction is obtained by the recalculation of image coordinates [102].

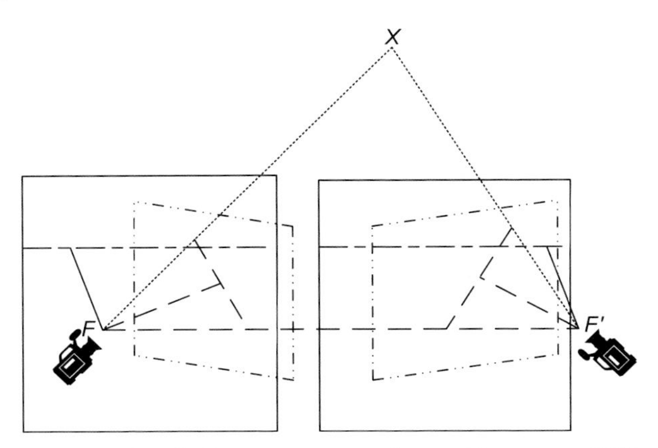


**Figure 52** Rectified images to support the matching process [102]

In Figure 52 the images before the recalculation are represented with frames, which are cyclically interrupted by two dots. The optical centers of the images are $F$ and $F$. The dashed epipolar lines belonging to these images are not parallel, in contrast to the lines, which are recurrently interrupted by a dot also representing epipolar lines, but belong to the two images represented by solid frames, which are the recalculated images after the adjustment. The adjusted coordinates are now symbolized by $\breve{u}$. For the adjustment two $3 \times 3$ transformation matrices are searched for to fulfill two equations: $\breve{u} = M\, u$ and $\breve{u}' = N\, u'$. As shown before, the epipoles of the adjusted images now move to infinity. The values for the matrices $M$ and $N$ must be determined. The adjustment performed with matrices $M$ and $N$ will not alter the two principal axes. This simplifies the search, because constraints diminish the number of unknowns. Here are formulas for the image coordinates after adjustment [102]:

$$\breve{u} = \begin{bmatrix} x_A \\ y_A \\ z_A \end{bmatrix} = M \begin{bmatrix} x_A \\ y_A \\ 1 \end{bmatrix}, \qquad \breve{u}' = \begin{bmatrix} x'_A \\ y'_A \\ z'_A \end{bmatrix} = N \begin{bmatrix} x'_A \\ y'_A \\ 1 \end{bmatrix}. \tag{6.33}$$

The calculation of the matrices $M$ and $N$ can be executed by the use of projection matrix $D$ described in formula (7.1). Because there are two images, two projection matrices $D$ and $D'$ exist before recalculation is performed. The three rows of the $3 \times 3$ submatrix on the left-hand side from $D$ are symbolized by $v_1$, $v_2$, and $v_3$. Analogously the symbols for the submatrix of $D$ belonging to the image on the right-hand side are $v'_1$, $v'_2$, and $v'_3$. The coordinates of the two optical centers are denoted by $F$ and $F'$. Now the two matrices $M$ and $N$ are written as [102]:

$$M = \begin{bmatrix} ((F \times F') \times F)^{\mathrm{T}} \\ (F \times F')^{\mathrm{T}} \\ ((F - F') \times (F \times F'))^{\mathrm{T}} \end{bmatrix} [v_2 \times v_3,\ v_3 \times v_1,\ v_1 \times v_2], \tag{6.34}$$

$$N = \begin{bmatrix} ((F \times F') \times F')^{\mathrm{T}} \\ (F \times F')^{\mathrm{T}} \\ ((F - F') \times (F \times F'))^{\mathrm{T}} \end{bmatrix} [v_2' \times v_3',\ v_3' \times v_1',\ v_1' \times v_2']. \qquad (6.35)$$

## 6.6 Ego-motion Estimation

We now consider the case when a calibrated camera is moved, but the ego-motion of the camera is unknown. So the movement can be described by rotation $T$ and translation $\tau$. These values must be examined from corresponding pixels in two images. These pixels will be denoted by $u_i$ and $u_i'$. An algorithm for the determination of the parameters $\tau$ and $T$ is now given[63, 100]:

1. A fundamental matrix must be derived. So corresponding points $u_i$ and $u_i'$ have to be examined.
2. Normalization can be helpful to diminish numerical errors.

$$\breve{u} = M_1 u, \quad \breve{u}' = M_2 u', \qquad (6.36)$$

$$M_1 = \begin{bmatrix} c_1 & 0 & c_3 \\ 0 & c_2 & c_4 \\ 0 & 0 & 1 \end{bmatrix}, \quad M_2 = \begin{bmatrix} c_1' & 0 & c_3' \\ 0 & c_2' & c_4' \\ 0 & 0 & 1 \end{bmatrix}. \qquad (6.37)$$

3. Now the fundamental matrix $\hat{B}$ must be estimated as explained in Section 6.4.
4. It is now possible to determine the essential matrix $\hat{E}$, because the calibration matrices $\Psi$ and $\Psi'$ are known:

$$\hat{E} = \Psi^{\mathrm{T}} \hat{B}\, \Psi'. \qquad (6.38)$$

5. The parameters $T$ and $\tau$ can be calculated with SVD by using the essential matrix $\hat{E}$:

$$\hat{E} = M\, Z\, N^{\mathrm{T}}, \quad Z = \begin{bmatrix} c_1'' & 0 & 0 \\ 0 & c_2'' & 0 \\ 0 & 0 & c_3'' \end{bmatrix}. \qquad (6.39)$$

As shown before, the diagonal matrix $Z$ should have two equal values, and the remaining coefficients are zero. But because of numerical errors two different values in $Z$ are expected. To obtain an appropriate adaptation of $Z$, $c_3$ should be set to zero and the arithmetic mean of $c_1$ and $c_2$ should be used:

$$E = M \begin{bmatrix} \frac{c_1'' + c_2''}{2} & 0 & 0 \\ 0 & \frac{c_1'' + c_2''}{2} & 0 \\ 0 & 0 & 0 \end{bmatrix} N^{\mathrm{T}}. \tag{6.40}$$

Finally, the determination of $T$ and $\tau$ by the decomposition of $E$ is carried out as shown in Section 6.3.

## 6.7
## Three-dimensional Reconstruction by Known Internal Parameters

We now consider the case when the accompanying three-dimensional coordinates of a scene point $X$ are estimated from two corresponding pixels $u$ and $u'$. But only the internal parameters of the two cameras are known. This means that the calibration matrices $\Psi$ and $\Psi'$ are known, but no knowledge about the external parameters exists. The formulas for $u$ and $u'$ are:

$$u \cong [\Psi|0]\ X, \quad u' \cong [\Psi' T| - \Psi' T\tau]\ X. \tag{6.41}$$

The algorithm for this problem can be applied in four steps:

1. Search for analogies,
2. The essential matrix $E$ has to be computed,
3. Rotation $T$and translation $\tau$ are derived from the essential matrix $E$,
4. $X$ must be calculated by using Equations (6.41).

It is not possible to get a full Euclidean reconstruction with this approach because of the lack of information about the distance between the cameras. Only a similarity reconstruction is available.

## 6.8
## Three-dimensional Reconstruction by Unknown Internal and External Parameters

### 6.8.1
### Three-dimensional Reconstruction with Two Uncalibrated Cameras

It is now assumed that no information exists about the internal and external parameters of the camera. Hence, both cameras are uncalibrated. No information about the camera position and the focal length used is given. The projection matrix $D$ is used to express the first camera's perspective. The matrix is composed of three row vectors $v_1^{\mathrm{T}}, v_2^{\mathrm{T}}, v_3^{\mathrm{T}}$ as shown in Equation (6.5). Apostrophes are used for the second camera:

First image: $$u = \begin{bmatrix} x_A \\ y_A \\ z_A \end{bmatrix} \cong D\,X = \begin{bmatrix} v_1^T \\ v_2^T \\ v_3^T \end{bmatrix} X, \tag{6.42}$$

Second image: $$u' = \begin{bmatrix} x'_A \\ y'_A \\ z'_A \end{bmatrix} \cong D'\,X = \begin{bmatrix} v'^T_1 \\ v'^T_2 \\ v'^T_3 \end{bmatrix} X. \tag{6.43}$$

No information about scaling is given. Therefore, this factor must be eliminated from $D$ by the consideration of the following relations:

$$x_A : y_A : z_A = v_1^T X : v_2^T X : v_3^T X, \tag{6.44}$$

$$x'_A : y'_A : z'_A = v'^T_1 X : v'^T_2 X : v'^T_3 X. \tag{6.45}$$

Hence, the following equations can be written:

$$\begin{array}{ll} x_A v_2^T X = y_A v_1^T X & x'_A v'^T_2 X = y'_A v'^T_1 X \\ x_A v_3^T X = z_A v_1^T X, & x'_A v'^T_3 X = z'_A v'^T_1 X. \\ y_A v_3^T X = z_A v_2^T X & y'_A v'^T_3 X = z'_A v'^T_2 X \end{array} \tag{6.46}$$

Now the matrix notation of Equation (6.46) for the first camera follows. The matrix for the second camera can be written analogously by using apostrophes:

$$\begin{bmatrix} x_A v_2^T - y_A v_1^T \\ x_A v_3^T - z_A v_1^T \\ y_A v_3^T - z_A v_2^T \end{bmatrix} X = 0. \tag{6.47}$$

The first and second rows of the matrix are multiplied by $z_A$ and $-y_A$, respectively, and then added. This is expressed in the next equation:

$$(x_A z_A v_2^T - y_A z_A v_1^T - x_A y_A v_3^T + y_A z_A v_1^T)X = (x_A z_A v_2^T - x_A y_A v_3^T)X = 0. \tag{6.48}$$

The next equation is constructed by the extraction of the third row from the matrix (6.47):

$$(-z_A v_2^T + y_A v_3^T)X = 0. \tag{6.49}$$

Formulas (6.48) and (6.49) are linearly dependent. This holds also for the second camera. Therefore, it is sufficient to use only two equations. Equations two and three are used in the next formula:

$$\begin{array}{ll}(x_A v_3^T - z_A v_1^T)\, X = 0 & (x'_A v'^T_3 - z'_A v'^T_1)\, X = 0 \\ (y_A v_3^T - z_A v_2^T)\, X = 0 , & (y'_A v'^T_3 - z'_A v'^T_2)\, X = 0 \end{array} . \tag{6.50}$$

The equations are now summarized in a matrix:

$$\begin{bmatrix} x_A v_3^T - z_A v_1^T \\ y_A v_3^T - z_A v_2^T \\ x'_A v'^T_3 - z'_A v'^T_1 \\ y'_A v'^T_3 - z'_A v'^T_2 \end{bmatrix} X = M\, X = 0. \tag{6.51}$$

$M$ is of size $4\times4$ and $X$ of size $4\times1$. To find a nontrivial solution for Equation (6.51), the case $\det(M) = 0$ is examined. This holds if the matrix has rank 3 provided that both pixels $u$ and $u'$ are indeed matching pixels. The further procedure depends now on the actual knowledge about the cameras. When the cameras are calibrated the known parameters are summarized in matrix $M$. $D$ and $D'$ are known just as $u$ and $u'$. So Equation (6.51) can be resolved. The other case considers uncalibrated cameras. In this case the examined three-dimensional point $\tilde{X}$ differs from the Euclidean reconstruction. The difference is expressed in an unknown matrix $N$ that provides a projective transformation from $X$ to $\tilde{X}$:

$$\tilde{X} = NX. \tag{6.52}$$

The dimension of the matrix $N$ is $4\times4$. $N$ is always the same for all scene points. $N$ changes if the camera position or the camera calibration changes:

$$u = \begin{bmatrix} x_A \\ y_A \\ z_A \end{bmatrix} \cong DX = DN^{-1}NX = \tilde{D}\tilde{X}, \tag{6.53}$$

$$u' = \begin{bmatrix} x'_A \\ y'_A \\ z'_A \end{bmatrix} \cong D'X = D'N^{-1}NX = \tilde{D}'\tilde{X}. \tag{6.54}$$

The formula shows that $DX$ and $\tilde{D}\tilde{X}$ are equal but $X$ and $\tilde{X}$ differ.

### 6.8.2 Three-dimensional Reconstruction with Three or More Cameras

It is now assumed that three or more cameras observe the same scene. It must be preconditioned that matching pixels exist in all taken images. Consider Figure 53.

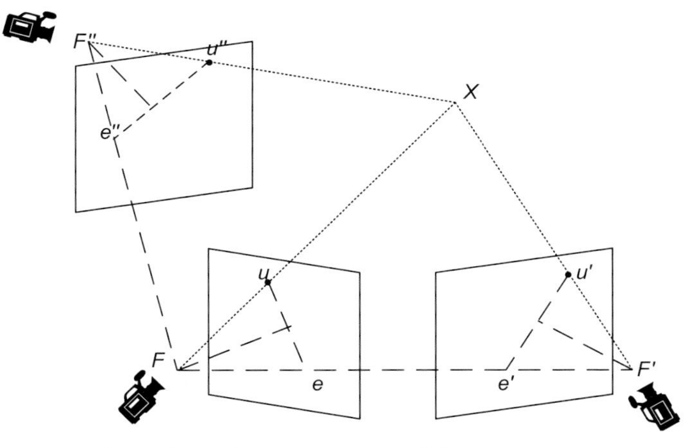


**Figure 53** Scene observed from three cameras [63]

Three cameras are used in the figure. The matching pixels in the images are denoted by $u$, $u'$, and $u''$. The accompanying three-dimensional points $X$, $X'$, and $X''$ are given by the projection matrices $D$, $D'$, and $D''$. The approach to derive three-dimensional points is based on the procedure for two uncalibrated cameras:

$$u = \begin{bmatrix} x_A \\ y_A \\ z_A \end{bmatrix} \cong DX = \begin{bmatrix} v_1^T \\ v_2^T \\ v_3^T \end{bmatrix} X, \quad u' = \begin{bmatrix} x'_A \\ y'_A \\ z'_A \end{bmatrix} \cong D'X = \begin{bmatrix} v'^T_1 \\ v'^T_2 \\ v'^T_3 \end{bmatrix} X, \tag{6.55}$$

$$u'' = \begin{bmatrix} x''_A \\ y''_A \\ z''_A \end{bmatrix} \cong D''X = \begin{bmatrix} v''^T_1 \\ v''^T_2 \\ v''^T_3 \end{bmatrix} X. \tag{6.56}$$

Now the unknown scaling factor must be eliminated. This can be executed in a similar manner as explained with formulas (6.43)–(6.51). As a result of this procedure a matrix is processed:

$$\begin{matrix} 1: \\ 2: \\ 3: \\ 4: \\ 5: \\ 6: \end{matrix} \begin{bmatrix} x_A v_3^T - z_A v_1^T \\ y_A v_3^T - z_A v_2^T \\ x'_A v'^T_3 - z'_A v'^T_1 \\ y'_A v'^T_3 - z'_A v'^T_2 \\ x''_A v''^T_3 - z''_A v''^T_1 \\ y''_A v''^T_3 - z''_A v''^T_2 \end{bmatrix} X = MX = 0. \tag{6.57}$$

The rows of the matrix are numbered and can be used for further reference. A nontrivial solution of the matrix (6.57) is to be found. This requires a matrix of rank 3. The matrix has rank 3 if the determinants of all $4 \times 4$ submatrices belonging to the matrix have the value 0. 15 submatrices exist. The submatrices must be clustered. It has to be determined if a submatrix involves two or three cameras. Three sets of equations can be built representing a bilinear relation between two cameras. The three sets are [1,2,3,4], [1,2,5,6], and [3,4,5,6]. In the case of three cameras there exists one scene point $X$ taken from different positions and appearing in three different images. This can be represented by a trilinear relation expressed by sets of equations. 12 trilinearities exist of which four are linearly independent. Three possibilities are shown for linearly independent $4 \times 4$ submatrices:

$$\begin{array}{llll} [1,2,3,5] & [1,2,4,5] & [1,2,3,6] & [1,2,4,6], \\ [1,3,4,5] & [2,3,4,5] & [1,3,4,6] & [2,3,4,6], \\ [1,3,5,6] & [1,4,5,6] & [2,3,5,6] & [2,4,5,6]. \end{array} \tag{6.58}$$

A row in the matrix (6.57) represents a plane in which the optical center $F$ and the scene point $X$ can be found, see Figure 54.

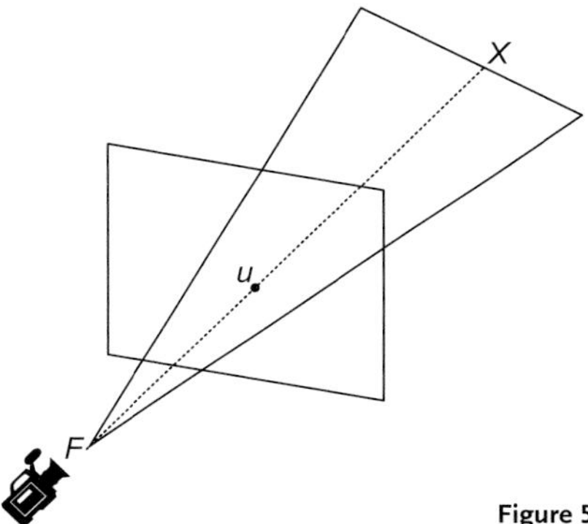


**Figure 54** Plane with optical center $F$ and scene point $X$ [63]

But with one trilinearity relation it can not be guaranteed that the three points $u$, $u'$, and $u''$ taken from the cameras match to the scene point $X$, see Figure 55.

It is depicted in the figure that only one ray connects an optical center with the scene point $X_1$. The other two rays meet the points $X_2$ and $X_3$ nearby $X_1$. It is possible to reduce the space to a plane with these additional two views in which the scene point $X$ can be found. So it can be said that one trilinearity relation can guarantee that a common point exists for the ray and the two planes, which lies in three-dimensional projective space. The number of cameras is now incremented to four cameras. In this case eight rows constitute the matrix (6.57). The $4 \times 4$ subdeterminants contain one row generated by one camera. This is a quadrilinear constraint. The quadrilinear constraint can be processed by a linear combination of bilinear and trilinear constraints. So it must be a precondition that all the bilinear and trilinear constraints are satisfied. This shows that it is not possible to obtain additional information for the scene point $X$ using more than three cameras, on condition that proper measurements took place. This geometric description is now written in alge-

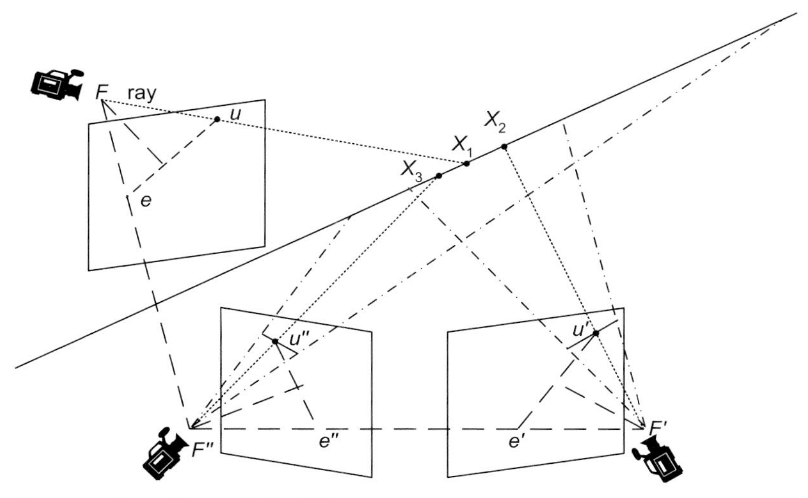


**Figure 55** One trilinear relation [63]

braic notation. It is now presumed that the first camera is in canonical configuration:

$$\begin{aligned} u &\cong D\tilde{X} = DN^{-1}NX = [Z|0]X \\ u' &\cong D'\tilde{X} = D'N^{-1}NX = [c'_{ij}]X \quad i = 1, \ldots, 3 \\ u'' &\cong D''\tilde{X} = D''N^{-1}NX = [c''_{ij}]X \quad j = 1, \ldots, 4 \end{aligned} \quad . \tag{6.59}$$

No knowledge is available about the scale in the image measurements $u \cong [Z|0]X$:

$$\tilde{X} = \begin{bmatrix} u \\ \rho \end{bmatrix}. \tag{6.60}$$

$\rho$ is the scale factor that must be determined as outlined before. The scene point $X$ is projected into the second camera with the next formula:

$$u' \cong [c'_{ij}]X = \begin{bmatrix} u \\ \rho \end{bmatrix}, \tag{6.61}$$

$$u'_i \cong c_i'^k u_k + c'_{i4}\rho \quad k = 1, \ldots, 3. \tag{6.62}$$

This is a short form without the summation symbol. The term $c_i'^k u_k$would be generally written as $\sum_{k=1}^{3} c'_{ik}u_k$. The scale factor must be extracted from this formula. The next formula shows three equations. Two of these three equations are independent:

$$u'_i(c_j'^k u_k + c'_{j4}\rho) = u'_j(c_i'^k u_k + c'_{i4}\rho). \tag{6.63}$$

These three equations are now resolved to $\rho$:

$$\rho = \frac{u_k(u_i' c_j'^k - u_j' c_i'^k)}{u_j' c_{i4}' - u_i' c_{j4}'}. \tag{6.64}$$

$\rho$ is eliminated by $\tilde{X}$:

$$\tilde{X} = \begin{bmatrix} u \\ \dfrac{u_k(u_i' c_j'^k - u_j' c_i'^k)}{u_j' c_{i4}' - u_i' c_{j4}'} \end{bmatrix} \cong \begin{bmatrix} (u_j' c_{i4}' - u_i' c_{j4}')u \\ u_k(u_i' c_j'^k - u_j' c_i'^k) \end{bmatrix}. \tag{6.65}$$

The projection of $\tilde{X}$ into the third camera occurs:

$$u'' \cong [c_{lk}'']X = c_l''^k \omega_k. \tag{6.66}$$

Some relations for $u_l''$ are:

$$\begin{aligned} u_l'' &\cong c_l''^k u_k(u_j' c_{i4}' - u_i' c_{j4}') + c_{l4}'' u_k(u_i' c_j'^k - u_j' c_i'^k) \\ &\cong u_k u_i (c_j'^k c_{l4}'' - c_{j4}' c_l''^k) - u_k u_j' (c_i'^k c_{l4}'' - c_{i4}' c_l''^k). \\ &\cong u_k(u_i' \Omega_{kjl} - u_j' \Omega_{kil}) \end{aligned} \tag{6.67}$$

A value that depends on three indices is called a tensor: $\Omega_{ijk}$, $i, j, k = 1, 2, 3$. A tensor is comparable to a three-dimensional matrix. It is possible to remove the unknown scale by joining all three views:

$$u_k(u_i' u_m'' \Omega_{kjl} - u_j' u_m'' \Omega_{kil}) = u_k(u_i' u_l'' \Omega_{kjm} - u_j' u_l'' \Omega_{kim}). \tag{6.68}$$

It is valid that $i < j$ and $l < m$. Nine equations can be observed. Four of these nine equations are linearly independent. It is now presumed that $j = m = 3$. For simplification it holds that $u_3 = u_3' = u_3'' = 1$. Further modifications are applied. This provides a trilinear constraint among three views:

$$u_k(u_i' u_l'' \Omega_{k33} - u_l'' \Omega_{ki3} - u_i' \Omega_{k3l} + \Omega_{kil}) = 0. \tag{6.69}$$

Indices $i$ and $l$ can be instantiated with value 1 or 2. Four linearly independent equations exist. 27 unknowns exist for the tensor $\Omega_{ijk}$. These unknowns can be estimated by using at least seven corresponding points in three images. Three advantages in association with trilinear constraints are given:

1. Seven corresponding points from three views are sufficient to derive the trilinear tensor. The examination of the fundamental matrix using a pair of views requires eight corresponding points.
2. Three fundamental matrices can be represented with the tensor.
3. It is expected that the estimation of the constraint will be numerically more stable if three views are used instead of three fundamental matrices.

## 6.9 Stereo Correspondence

The computation of correspondences between pixels in different views is a necessary precondition to obtain depth information. Pixels that represent the same in different images must be found and thereafter geometrically analyzed. The automatic detection of corresponding pixels in different images is an actual research topic. Probably, it is not possible to find an approach that provides a general solution. Rather, specific features, which are dependent on the computer-vision task, must appear in the image. These features will be used to find corresponding pixels. For example, imagine a black square that appears in both images, where it occupies most of the image area. The gray values representing the square are rather identical in both images. In this case it is not possible to find corresponding pixels lying inside the square area. It may also be the case that some areas of an object can not be seen from both cameras and appear therefore only in one image. But such occurrences are rather rare in practical applications.

Generally it is possible to find matching features in the images. To reduce ambiguities, it is recommended to use several features. For example, features can be derived from geometry, which affects the image taking, photometry, or from the object attributes. Many approaches to calculate correspondences between pixels have been developed. Some of them are outlined at this point [103]:

1. Epipolar constraint: To find the corresponding pixel in the second image its epipolar line is used, because the pixel must lie on this line. The search space is diminished from two-dimensional space to one-dimensional space.
2. Photometric compatibility constraint: The gray values of the pixels are used. It can be assumed that the gray values of corresponding pixels are nearly equal. They are probably not completely equal, because the luminosity differs due to different positions from which the images are taken.
3. Geometric similarity constraints: Geometric attributes of objects like length of lines, contours, or regions are used to get corresponding pixels. It is supposed that the attribute values are equal.

These three approaches exploit geometrical or photometrical characteristics. The following methods are based on object attributes:

1. Disparity smoothness constraint: This method is based on the assumption that the amount of disparity differences between adjacent pixels is similar in both images. Let $p_L^1$ and $p_L^2$ be the coordinates of two adjacent pixels in the left image. $p_R^1$ and $p_R^2$ are the corresponding coordinates in the right image, then the absolute difference computed with the following formula is small, on condition that the two images were taken from cameras arranged in parallel (canonical configuration):

$$||p_L^1 - p_R^1| - |p_L^2 - p_R^2||. \tag{6.70}$$

2. Figural continuity constraint: Additionally to the fulfillment of the disparity smoothness constraint it is required that the neighboring pixels lie on an edge in both images.
3. Disparity limit: In psychophysical experiments it has been verified that stereo images can only be merged if the disparity does not exceed a limit.
4. Ordering constraint: The corresponding pixels of objects that have similar depths have the same order on epipolar lines of both images. This is not valid if an object point is close to the camera and has additionally a large depth difference to other objects in the scene. In this case the corresponding pixels on epipolar lines can have a different sorting on both lines.

Also a strategy was suggested to detect pseudocorresponding pixels. A mutual correspondence constraint can be used for these purposes. Pseudocorrespondence can result from noise or shadow cast. It shall be assumed that the search started from the left image. If two (probably) corresponding pixels $p_L$ and $p_R$ have been found, the result is checked. Therefore, a second search starts from the right image. If pseudocorrespondence exists, the second search will probably generate another result. The two pixels do not match.

### 6.9.1 Correlation-based Stereo Correspondence

Correlation-based stereo correspondence uses the gray values to find matching pixels in two images. Indeed analyzing only pixels is not sufficient, because there may be several candidates that have the same gray value. It is necessary to examine the neighborhood. For example, this happens with windows of sizes $5 \times 5$ or $8 \times 8$. Block matching is an approach that considers the neighborhood to find stereo correspondence. It shall be presumed that the canonical configuration for two cameras is used. Block matching tries to find matching blocks in both images. In these matching blocks only one disparity exists between the pixels' gray values for each block of the two images. Block matching starts with the decomposition of one of the two images into blocks. This may be the right image. Then it tries to find corresponding blocks in the left image. Statistical measurements for the gray values are used to check the correspondence between two blocks. At the end of the block matching a matrix should exist that contains for every block a particular disparity that correlates to a representative pixel in the block [103].

A drawback of the block-matching method is the relatively long runtime, which is in general a symptom of stereo-correspondence algorithms that are trying detections of correlations between pixels.

### 6.9.2 Feature-based Stereo Correspondence

Approaches that belong to the class of feature-based stereo correspondence, search for conspicuous criteria like edges, pixels along edges, and so on. To detect correspondence, the attributes of these features are also used. For example, these can be the edge length, the orientation, and others. Approaches that are feature-based have some advantages in comparison to intensity-based approaches [103]:

1. Ambiguity can be diminished, because the number of pixels that must be examined is rather small.
2. The calculation is not so strongly biased by photometric confusions like inhomogeneous illumination.
3. The calculation of disparities can be executed with subpixel precision.

The PMF algorithm is a feature-based method. Prerequisite for this method is the detection of a set that contains pixels that fulfill a feature. These pixels must be extracted from the images with appropriate operators. For example, operators for edge detection can be used. The algorithm obtains these pixels as input and processes pixel pairs in both images, which belong together. Three constraints are used to find corresponding pixels. These are the epipolar constraint, the uniqueness constraint, and the disparity gradient limit constraint. The uniqueness constraint presumes that generally a pixel from the first image can only be associated to one pixel in the second image. The disparity gradient is used to calculate the relative disparity between two matching pixel pairs. Consider Figure 56.

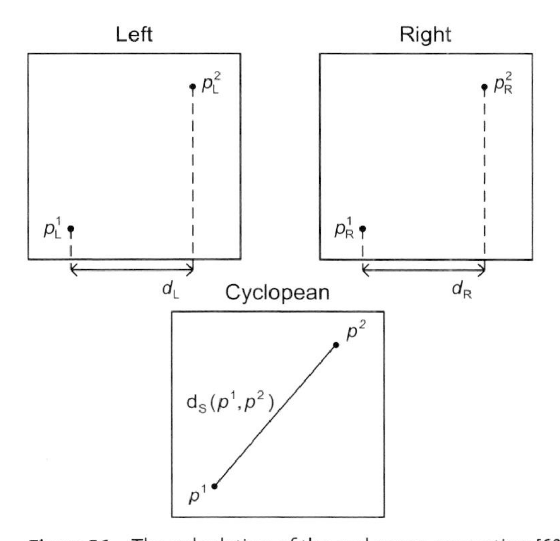


**Figure 56** The calculation of the cyclopean separation [63]

In the left and right images two pixels $p^1$ and $p^2$ can be found, which have been taken from a three-dimensional scene. The coordinates of $p_L^1$ in the left image are $p_L^1 = (x_{A_L}^1, y_A^1)$ and of $p_L^2 = (x_{A_L}^2, y_A^2)$. The coordinates for the two pixels in the right image are denoted similarly by $p_R^1 = (x_{A_R}^1, y_A^1)$ and $p_R^2 = (x_{A_R}^2, y_A^2)$. Note that the $y$ coordinates must be equal for the pixel pairs. This is a demand to fulfill the epipolar constraint. The cyclopean image is generated with the average of the $x$ coordinates in the left and right images:

$$p_{\mathrm{C}}^{1} = \left( \frac{x_{\mathrm{A_L}}^{1} + x_{\mathrm{A_R}}^{1}}{2}, y_{\mathrm{A}}^{1} \right), \tag{6.71}$$

$$p_{\mathrm{C}}^{2} = \left( \frac{x_{\mathrm{A_L}}^{2} + x_{\mathrm{A_R}}^{2}}{2}, y_{\mathrm{A}}^{2} \right). \tag{6.72}$$

The cyclopean separation $d_S$ can also be found in Figure 56 that represents the distance between the two pixels in the cyclopean image:

$$\begin{aligned} d_S(p^1, p^2) &= \sqrt{\left[ \left( \frac{x_{\mathrm{A_L}}^{1} + x_{\mathrm{A_R}}^{1}}{2} \right) - \left( \frac{x_{\mathrm{A_L}}^{2} + x_{\mathrm{A_R}}^{2}}{2} \right) \right]^2 + \left( y_{\mathrm{A}}^{1} - y_{\mathrm{A}}^{2} \right)^2} \\ &= \sqrt{\frac{1}{4} \left[ \left( x_{\mathrm{A_L}}^{1} - x_{\mathrm{A_L}}^{2} \right) + \left( x_{\mathrm{A_R}}^{1} - x_{\mathrm{A_R}}^{2} \right) \right]^2 + \left( y_{\mathrm{A}}^{1} - y_{\mathrm{A}}^{2} \right)^2} \\ &= \sqrt{\frac{1}{4} (d_{\mathrm{L}} + d_{\mathrm{R}})^2 + \left( y_{\mathrm{A}}^{1} - y_{\mathrm{A}}^{2} \right)^2} \end{aligned}. \tag{6.73}$$

The next formula is used to process disparity between the pixel pair in the cyclopean image:

$$\begin{aligned} U(p^1, p^2) &= \left( x_{\mathrm{A_L}}^{1} - x_{\mathrm{A_R}}^{1} \right) - \left( x_{\mathrm{A_L}}^{2} - x_{\mathrm{A_R}}^{2} \right) \\ &= \left( x_{\mathrm{A_L}}^{1} - x_{\mathrm{A_L}}^{2} \right) - \left( x_{\mathrm{A_R}}^{1} - x_{\mathrm{A_R}}^{2} \right) \\ &= d_{\mathrm{L}} - d_{\mathrm{R}} \end{aligned}. \tag{6.74}$$

The disparity gradient is the ratio between disparity and cyclopean separation:

$$\begin{aligned} \Gamma(p^1, p^2) &= \frac{U(p^1, p^2)}{d_S(p^1, p^2)} \\ &= \frac{d_{\mathrm{L}} - d_{\mathrm{R}}}{\sqrt{\frac{1}{4} (d_{\mathrm{L}} + d_{\mathrm{R}})^2 + \left( y_{\mathrm{A}}^{1} - y_{\mathrm{A}}^{2} \right)^2}} \end{aligned}. \tag{6.75}$$

The disparity gradient will exceed the value one rather rarely in practical use. This constraint means that a short disparity between two scene points in the three-dimensional scene is not permitted if the points are very close together. A solution is then generated with a relaxation procedure. The possible matching points are valued. The evaluation uses the fixed boundary for the disparity gradient as a criterion. It is checked if further matches exist for the pixel pair that also meet the boundary. A match is considered as correct if it has a high valuation. The detected matches are used in a downstream process to find further matches. Six steps are now listed that explain the PMF algorithm:

1. Features like edge pixels must be determined that can be used for stereo correspondence.
2. One of the two images must be chosen to find pixels for stereo correspondence. These pixels are used to find the associated pixels in the other image with epipolar lines.
3. The score of the matching pixels must be raised. The actual value depends on the matches that have been found before. It is additionally required that these pixels may not exceed the determined disparity gradient limit.
4. The match with the highest score for both pixels is accepted. Further examinations for this pixel pair are not performed to fulfill the uniqueness constraint.
5. The scores must be calculated again. The algorithm terminates if all possible matches have been found. Otherwise it must be continued with step two.

The using of epipolar lines in step two reduces the search space to one dimension and the uniqueness constraint assures that a pixel is only used once for the calculation of the disparity gradient.

The PMF algorithm can ran very fast. Its procedure is suitable for parallel implementation. It is difficult to find corresponding pixels in horizontal lines with PMF, because these lines extend often over neighboring grids that aggravates the detection with epipolar lines.

## 6.10 Image-sequence Analysis

The evaluation of temporal image sequences belongs to the passive optical sensor methods. Image-sequence analysis can extract three-dimensional information without a priori knowledge of the objects contained in the image. The image-sequence analysis can be used to determine a relative three-dimensional movement between the camera and the scene. For example, this scenario can be found by autonomous robots. If the movement is known, it is possible to determine the three-dimensional information with the image-sequence analysis by using only one camera, which is also called pseudostereo vision. The most general task in image-sequence analysis is the determination of the three-dimensional structure of a scene and the examination of the parameters that depict the relative movement between the camera and the scene. The first task in image-sequence analysis is the determination of an association between two successive images. The matching of the images can be realized on the basis of optical flow. The optical flow is represented in a vector field in which the difference of the successive images is tracked relating to time. In most instances the difference of the pixels between the two successive images can be found in the vector field, but it must be said that the tracking of pixel differences in the vector field is very time consuming. The optical flow is used to process a depth map. The depth map is generated by using a smoothness condition, but it can be that disturbances effect confusion by applying the smoothness condition. To guarantee a proper depth map, a dense scanning in a time slice is required. This will ensure that the pixel differences between the two images are rather small. Therefore, the methods based on the optical flow are often called continuous approaches. A further precondition for the approaches based on the optical flow is the constancy of the illumination between the images. But in reality this is not often true. Feature extraction is an alternative approach that is independent of the alteration of the illumination. Points or lines must be used generally for the feature extraction. Otherwise additional restrictions must be defined. The feature-extraction approach offers the possibility for larger lags between the taking of two successive images. The independence from changes in the illumination provides a more robust construction of the depth map. First, the depth map will be rather sparsely populated, because the extracted features are distributed with quite wide distances in the image. But this approach also has drawbacks. The features can be temporarily occluded or image data can be disturbed that make it harder to generate the depth map. A discontinuous movement is also a problem here. Approaches exist that use a system of equations to process the unknown movement and/or structure parameters. The problem that is to be solved determines if a linear or nonlinear system of equations must be used. A linear system of equations can be solved with well-known mathematical algorithms, for example, the Gauss–Jordan algorithm. Iterative solution methods can be used for nonlinear systems of equations [93].

## 6.11 Three-dimensional Reconstruction from Image Sequences with the Kalman Filter

An application is now depicted in which the three-dimensional structure of an object is calculated, whereby the movement parameters, which are relative to the camera, are known. The three-dimensional reconstruction is based on the analysis of two-dimensional image sequences. This is known as a dynamic estimation problem with a nonlinear observation equation. Parameters are required for the three-dimensional geometrical representation. These parameters are three-dimensional world coordinates $(x, y, z)_{W_k}$ at a point in time $k$. The taking of the image sequence is a representation of the system in discrete time. The projection matrix $D_k$ describes the mapping of three-dimensional world coordinates to two-dimensional sensor coordinates $(x, y)_{S_k}$ at a point in time $k$. Estimation approaches for nonlinear dynamic systems can be realized with a linear system and/or an observation equation in the environment of an operating point. The linear equations can be derived with a Taylor series. For these purposes an extended Kalman filter can be used that is a suboptimal estimation approach. The estimation of a state vector is only an approximation for the estimated value of the minimal variance and the covariance matrix of the estimated error. Therefore, a test is required to obtain hints whether the extended Kalman filter provides good estimated values. The equations of the discrete extended Kalman filter are now shown [93]:

$$\text{The prediction of the state vector: } (x, y, z)^{-}_{\hat{W}_{k+1}} = Z_k \cdot (x, y, z)^{+}_{\hat{W}_k} \,. \tag{6.76}$$

The prediction of the covariance matrix: $P_{k+1}^{-} = Z_k \cdot P_k^{+} \cdot Z_k^{\mathrm{T}} \cdot Q_k.$ (6.77)

The update of the state vector:

$$(x, y, z)^{+}_{\hat{\mathrm{W}}_k} = (x, y, z)^{-}_{\hat{\mathrm{W}}_k} + K_k \cdot [(x, y)_{\mathrm{S}_k} - D_k \cdot (x, y, z)^{-}_{\hat{\mathrm{W}}_k}]. \tag{6.78}$$

The update of the covariance matrix:

$$P_k^{+} = [N - K_k \cdot L_k \cdot (x, y, z)^{-}_{\hat{\mathrm{W}}_k}] \cdot P_k^{-}. \tag{6.79}$$

Kalman enhancement:

$$K_k = P_k^{-} \cdot L_k^{\mathrm{T}} \cdot (x, y, z)^{-}_{\hat{\mathrm{W}}_k} \cdot [L_k \cdot (x, y, z)^{-}_{\hat{\mathrm{W}}_k} \cdot P_k^{-} \cdot L_k^{\mathrm{T}} \cdot (x, y, z)^{-}_{\hat{\mathrm{W}}_k} + M_k]^{-1}. \tag{6.80}$$

Figure 57 shows a state-space representation of the system model and the Kalman filter.

The predicted state vector is used as the operating point. The values of the Kalman enhancement matrix $K_k$ are random variables. The values depend on the Jacobi matrix and the state vector $L_k \cdot (x, y, z)^{-}_{\hat{\mathrm{W}}_k}$. The series of enhancement matrices must be calculated during the runtime. The construction of the Kalman filter depends on two assumptions. The image coordinates must be determinable from the three-dimensional world coordinates. This is possible by performing the calibration procedure. Parameters that represent the relative movement between the camera and the object, must be known [93].

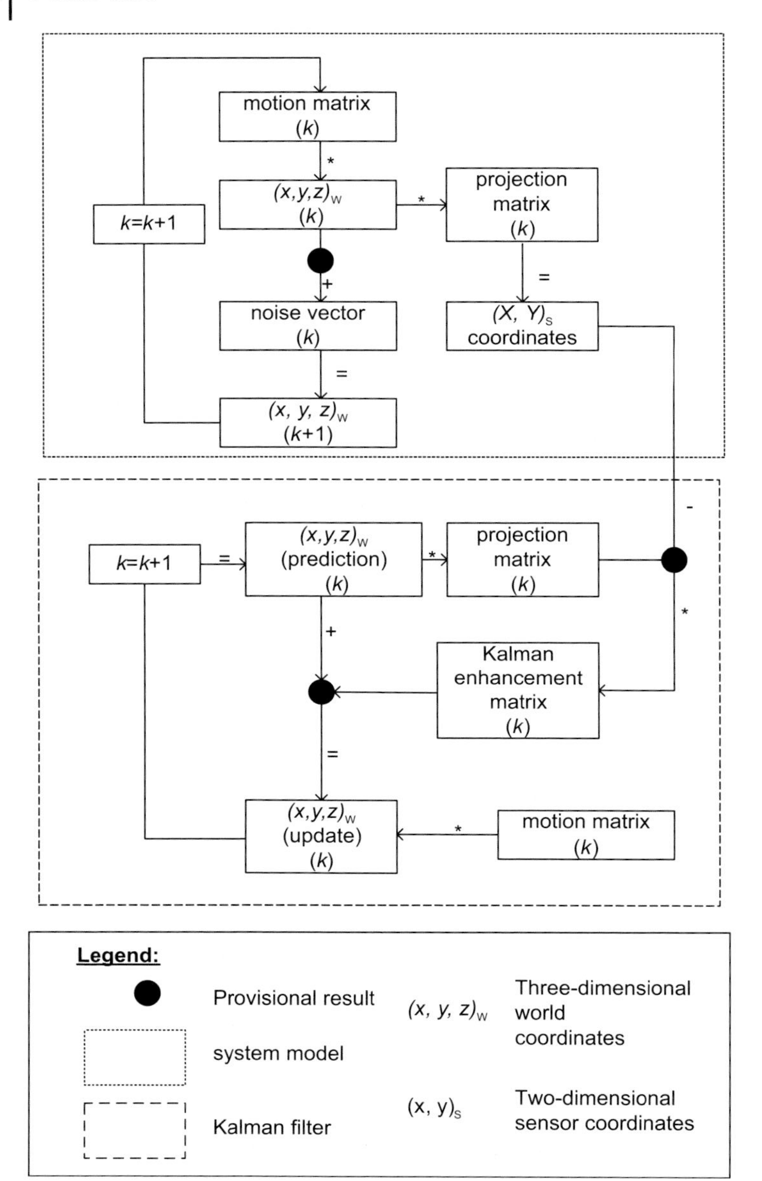


**Figure 57** State-space representation [93]

# 7
# Camera Calibration

Calibration is the determination of the camera model, for example, a pinhole-camera model, parameters. A division into two classes of camera-calibration approaches can be observed. Approaches that belong to the first class [104, 105], are based on a physical camera model like the pinhole-camera model. The approaches of the second class [106, 107] are based only on mathematical models.

Figure 58 shows a robot-vision system with coordinate systems.

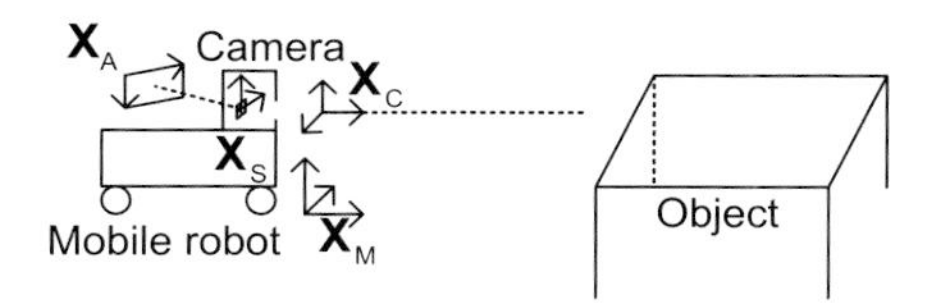


**Figure 58** Coordinate systems of a robot-vision system [33]

Two calibrations are necessary. The camera and the robot must be calibrated. The process of camera calibration provides internal camera parameters. The relative position of a camera coordinate system $\mathbf{X}_C$ to another coordinate system, for example, the robot coordinate system $\mathbf{X}_M$ is specified by external parameters. The camera calibration will enable determination of parameters that explain the projection of a three-dimensional object to the two-dimensional plane. A precise camera calibration permits an exact three-dimensional object reconstruction by using information about the object, such as size, position, and movement. The application determines the necessary precision of the calibration method. The worse the camera projection the more exact must be the calibration process. Many calibration approaches exist that calculate the internal and the external camera parameters. Two categories of calibration approaches exist. These are test-area-calibration approaches and self-calibration approaches. Test-area-calibration approaches use images where the three-dimensional world coordinates are known to derive the internal and external camera parameters. Very precisely measured test areas or reference objects are often used. The manufacturing and measuring of such reference objects requires much effort and is often afflicted with errors. The self-calibration approaches determine the external and internal camera parameters as well as the world coordinates of the ref-

*Robot Vision: Video-based Indoor Exploration with Autonomous and Mobile Robots.* Stefan Florczyk

ISBN: 3-527-40544-5

erence points. For these purposes several images of an unmeasured reference object taken from different camera positions are necessary. In the majority of cases the self-calibration procedure needs appropriate start values to determine the unknown parameters in an iterative process. Self-calibration approaches can be further divided into passive and active self-calibration approaches. Passive self-calibration approaches need no knowledge about the movement of the camera in contrast to active self-calibration approaches [33].

The next sections describe the calibration of a pinhole camera according to [63] and the self-calibration of a robot system that is described in [33]:

## 7.1 The Calibration of One Camera from a Known Scene

### 7.1.1 Pinhole-camera Calibration

The calibration of one camera from a known scene often requires two steps. The estimation of the projection matrix $D$ takes place in the first step. Therefore, it is necessary to use the known world coordinates in the scene. The internal and external parameters are calculated in the second step. To derive the projection matrix, equations are used that show the relationship between a point $X = (x, y, z)_{\mathrm{W}}^{\mathrm{T}}$ in the world coordinate system and its corresponding point in the image affine coordinate system $(x, y)_{\mathrm{A}}^{\mathrm{T}}$. Unknown values $c_{ij}$ can be found in a $3 \times 4$ projection matrix that must be calculated:

$$\begin{pmatrix} \rho x_{\mathrm{A}} \\ \rho y_{\mathrm{A}} \\ \rho \end{pmatrix} = \begin{pmatrix} c_{11} + c_{12} + c_{13} + c_{14} \\ c_{21} + c_{22} + c_{23} + c_{24} \\ c_{31} + c_{32} + c_{33} + c_{34} \end{pmatrix} \begin{pmatrix} x_{\mathrm{W}} \\ y_{\mathrm{W}} \\ z_{\mathrm{W}} \\ 1 \end{pmatrix}, \tag{7.1}$$

$$\begin{pmatrix} \rho x_{\mathrm{A}} \\ \rho y_{\mathrm{A}} \\ \rho \end{pmatrix} = \begin{pmatrix} c_{11} x_{\mathrm{W}} + c_{12} y_{\mathrm{W}} + c_{13} z_{\mathrm{W}} + c_{14} \\ c_{21} x_{\mathrm{W}} + c_{22} y_{\mathrm{W}} + c_{23} z_{\mathrm{W}} + c_{24} \\ c_{31} x_{\mathrm{W}} + c_{32} y_{\mathrm{W}} + c_{33} z_{\mathrm{W}} + c_{34} \end{pmatrix}, \tag{7.2}$$

$$\begin{matrix} x_{\mathrm{A}} (c_{31} x_{\mathrm{W}} + c_{32} y_{\mathrm{W}} + c_{33} z_{\mathrm{W}} + c_{34}) = c_{11} x_{\mathrm{W}} + c_{12} y_{\mathrm{W}} + c_{13} z_{\mathrm{W}} + c_{14} \\ y_{\mathrm{A}} (c_{31} x_{\mathrm{W}} + c_{32} y_{\mathrm{W}} + c_{33} z_{\mathrm{W}} + c_{34}) = c_{21} x_{\mathrm{W}} + c_{22} y_{\mathrm{W}} + c_{23} z_{\mathrm{W}} + c_{24} \end{matrix} . \tag{7.3}$$

Two linear equations are available for a known point in the world coordinate system and its corresponding point in the image coordinate system. If $n$ points are examined, the equations can be written as a $2n \times 12$ matrix with 12 unknowns:

$$\begin{pmatrix} x_{\mathrm{W}} & y_{\mathrm{W}} & z_{\mathrm{W}} & 1 & 0 & 0 & 0 & 0 & -x_{\mathrm{A}} x_{\mathrm{W}} & -x_{\mathrm{A}} y_{\mathrm{W}} & -x_{\mathrm{A}} z_{\mathrm{W}} & -x_{\mathrm{A}} \\ 0 & 0 & 0 & 0 & x_{\mathrm{W}} & y_{\mathrm{W}} & z_{\mathrm{W}} & 1 & -y_{\mathrm{A}} x_{\mathrm{W}} & -y_{\mathrm{A}} y_{\mathrm{W}} & -y_{\mathrm{A}} z_{\mathrm{W}} & -y_{\mathrm{A}} \\ & & & & & & \vdots & & & & & \end{pmatrix} \begin{pmatrix} c_{11} \\ c_{12} \\ \vdots \\ c_{34} \end{pmatrix} = 0. \tag{7.4}$$

Eleven unknown parameters can be observed in the matrix $D$ if homogeneous coordinates are used. Six corresponding scene and image points at least are necessary to get a solution. Additional points are often used. So the least-squares method can be used to solve the overdetermined Equation (7.4). This procedure has the advantage that errors in the measurements can be eliminated. The projection matrix $D$ is computed as a result:

$$D = [\Psi T | -\Psi T \tau] = [M | \nu]. \tag{7.5}$$

External parameters, which are the rotation $T$ and the translation $\tau$, must be extracted from the projection matrix. $M$ is a $3 \times 3$ submatrix from $D$ and $\nu$ the last column on the right-hand side.

$\Psi$ is called the camera-calibration matrix:

$$\tilde{u} = \begin{pmatrix} x_A \\ y_A \\ z_A \end{pmatrix} = \begin{pmatrix} c_1 & c_2 & -x_A^0 \\ 0 & c_3 & -y_A^0 \\ 0 & 0 & 1 \end{pmatrix} \begin{pmatrix} \frac{-b x_C}{z_C} \\ \frac{-b y_C}{z_C} \\ 1 \end{pmatrix} = \begin{pmatrix} -b\, c_1 & -b\, c_2 & -x_A^0 \\ 0 & -b\, c_3 & -y_A^0 \\ 0 & 0 & 1 \end{pmatrix} \begin{pmatrix} \frac{x_C}{z_C} \\ \frac{y_C}{z_C} \\ 1 \end{pmatrix}. \tag{7.6}$$

$\tilde{u}$ represents a computed two-dimensional point in the image plane I in homogeneous coordinates derived from the projection of the corresponding three-dimensional camera coordinates $(x, y, z)_C$, see Figure 59.

The next equation shows the relation between $u_C$ and the corresponding three-dimensional camera coordinates, where $b$ is the focal length:

$$u_C = \left( \frac{-b x_C}{z_C}, \quad \frac{-b y_C}{z_C}, \quad -b \right)^T. \tag{7.7}$$

Point $p^0$ represents the intersection of the optical axis with the image plane I: $p_A^0 = (x_A^0, \ y_A^0, \ 0)^T$. The point $u$ can be written in homogeneous coordinates as $\tilde{u} = (x_A, \ y_A, \ z_A)^T$. $X_A$, $Y_A$, and $Z_A$ are the coordinate axes of the image affine coordinate system $\mathbf{X}_A$. It is possible to compute the affine transformation with a $3 \times 3$ matrix with homogeneous coordinates. The homogeneous matrix has the unknowns $c_1$, $c_2$, and $c_3$. The translation vector can be computed from the formula (7.5) by the substitution from $M$ with $\Psi T$ and resolution to

$$\tau = -M^{-1} \nu. \tag{7.8}$$

The determination of the calibration matrix $\Psi$ must consider the fact that $\Psi$ is upper triangular and the rotation matrix $T$ is orthogonal. $M$ can be split into two matrices to obtain $\Psi$ and $T$ with the matrix-factorization method [108]: QR-decomposition.

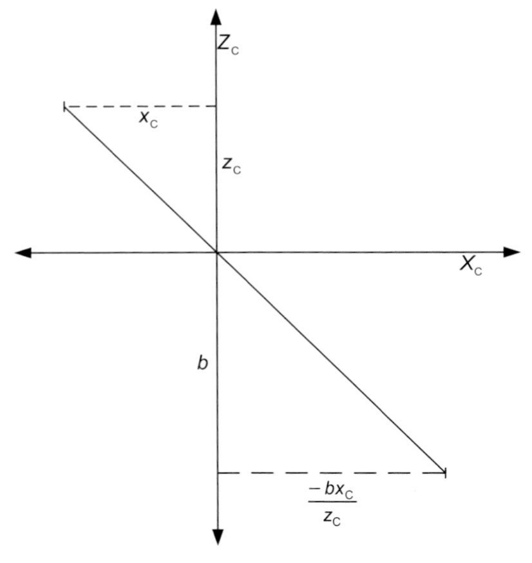


| Legend: | |
|---|---|
| $X_C$, $Z_C$ | Coordinate axes of the camera coordinate system |
| $b$ | Focal length |
| $x_C$, $z_C$ | Distances |

**Figure 59** Relation between the coordinates of the projected point [63]

### 7.1.2 The Determination of the Lens Distortion

To refine the camera calibration, the lens distortion can be added to the pinhole-camera model, because it is a feature of real cameras. Two kinds of distortion can be observed. Radial distortion bends the camera's line of sight and de-centering shifts the principal point from the principal axis. The next equations show five external parameters:

$$x_A = \frac{\mathbf{x}_A}{\mathbf{z}_A} = -b\,c_1\frac{x_C}{z_C} - b\,c_2\frac{y_C}{z_C} - x_A^0 = \alpha_{X_A}\frac{x_C}{z_C} + \alpha_{\text{shear}}\frac{y_C}{z_C} - x_A^0, \tag{7.9}$$

$$y_A = \frac{\mathbf{y}_A}{\mathbf{z}_A} = -b\,c_3\frac{y_C}{z_C} - y_A^0 = \alpha_{Y_A}\frac{y_C}{z_C} - y_A^0. \tag{7.10}$$

Substitutions $\alpha_{X_A} = -bc_1$, $\alpha_{\text{shear}} = -bc_2$, and $\alpha_{Y_A} = -bc_3$ are executed in these formulas. The internal parameters have the unit pixels. Measuring $b$ with the $X_A$-axis scaling is represented by $\alpha_{X_A}$. $b$ has the unit pixel. If $b$ is measured in pixels with the $Y_A$-axis scaling, the symbol $\alpha_{Y_A}$ is used. Obliquity between the $X_A$-axis of the image affine coordinate system in comparison to the $Y_I$-axis is measured with $\alpha_{\text{shear}}$. So $\alpha_{\text{shear}}$ also represents the obliquity between $b$ and $Y_I$, because the $X_A$-axis and $b$ are coincident. The equations show that the focal length is replaced by a camera constant. The camera constant has the same value as the focal length in the simple pinhole-camera model. This is not true for real cameras. In this case the coincidence holds only if the focus is set to infinity. If this is not the case, the camera constant has a lower value than the focal length. To determine these internal parameters, known calibration images with continuous patterns like points are used that are distributed on the whole image. So the internal parameters can be determined by the examination of distortions in the pattern. Radial and decentering distortions are considered often as rotationally symmetric and are modeled in many cases with polynomials. $x_A$ and $y_A$ are correct image coordinates in the following equations. $x$ and $y$ are pixel coordinates. $\tilde{x}_A$ and $\tilde{y}_A$ are the measured image coordinates, which are inexact. $\tilde{x}_A$ and $\tilde{y}_A$ are computed by using the pixel coordinates $x$, $y$ and the principal point coordinates $\hat{x}_A^0$ and $\hat{y}_A^0$:

$$\tilde{x}_A = x - \hat{x}_A^0, \tag{7.11}$$

$$\tilde{y}_A = y - \hat{y}_A^0. \tag{7.12}$$

To obtain the correct values of the image coordinates $\tilde{x}_A$ and $\tilde{y}_A$, it is necessary to add balances $\zeta_{x_A}$ and $\zeta_{y_A}$ to the inexact values $\tilde{x}_A$ and $\tilde{y}_A$:

$$x_A = \tilde{x}_A + \zeta_{x_A}, \tag{7.13}$$

$$y_A = \tilde{y}_A + \zeta_{y_A}. \tag{7.14}$$

To determine $\tilde{x}_A$ and $\tilde{y}_A$, polynomials of higher degree are often used to reflect the rotational symmetry:

$$\zeta_{x_A} = (\tilde{x}_A - \zeta_{x_A^0})(c_1\psi^2 + c_2\psi^4 + c_3\psi^6), \tag{7.15}$$

$$\zeta_{y_A} = (\tilde{y}_A - \zeta_{y_A^0})(c_1\psi^2 + c_2\psi^4 + c_3\psi^6). \tag{7.16}$$

$\zeta_{x_A^0}$ and $\zeta_{y_A^0}$ are balances for the shifting of the principal point. $\psi^2$ represents the square of the radial distance with regard to the center:

$$r^2 = (\tilde{x}_A - \zeta_{x_A^0}) + (\tilde{x}_A - \zeta_{x_A^0}). \tag{7.17}$$

The determination of the balances $\zeta_{x_A^0}$ and $\zeta_{y_A^0}$ is quite simple, because $\hat{x}_A^0$ and $\hat{y}_A^0$ can be used for the calculation:

$$x_A^0 = \hat{x}_A^0 + \zeta_{x_A^0}, \tag{7.18}$$

$$y_A^0 = \hat{y}_A^0 + \zeta_{y_A^0}. \tag{7.19}$$

To obtain a simplification of Equations (7.15) and (7.16), a polynomial of second order is often used. The simplification can be gained, for example, by omitting the principal point shifting:

$$x_A = \tilde{x}_A[1 \pm c_1((\tilde{x}_A)^2 + (\tilde{y}_A)^2)], \tag{7.20}$$

$$y_A = \tilde{y}_A[1 \pm c_1((\tilde{x}_A)^2 + (\tilde{y}_A)^2)]. \tag{7.21}$$

Two calibration methods are now outlined that are appropriate to determine the lens distortion, but require additional data in contrast to the pinhole-camera calibration.

HALCON offers a calibration approach. 10 to 20 images are required from this approach. The images can show a planar calibration table. The images of an object must be taken from diverse positions and orientations. It is further required that the object must fill at least a quarter of the image. HALCON's calibration operator also asks for the technical data of a camera. These are the nominal focal length $b$, the horizontal size $g_x$, and the vertical size $g_y$ of a cell on a CCD sensor. A unit meter must be used for these initial values. The central point of an image must be provided with coordinates $(x^c, y^c)_A$. The coordinates in pixels can be simply processed by the division of the image width and height by two [4].

Roger Tsai [109] has developed a calibration method that is common in industrial applications. The following explanation of the method is from [110]. The method needs the principal point $(x^0, y^0)_A$. But the scale factors $g_x$ and $g_y$ are also required and an aspect distortion factor $\psi_\rho$ is needed. $\psi_\rho$ represents the distortion with regard to the image width and image height. The method calculates the focal length $b$ and the radial distortion $\psi_D$. The method has similarities with HALCON's calibration operator. The method uses a calibration table that must be drawn on a sheet of paper and attached onto a vertical plate. The plate was joined with a rail that permitted the displacement of the plate in 10-mm steps. Therefore, the distance between camera and calibration object could be altered in well-defined and simply measurable steps.

## 7.2 Calibration of Cameras in Robot-vision Systems

In the following, two different self-calibration approaches are explained that were developed for cameras with a rigid focal length:

1. Calibration with moved calibration object,
2. Calibration with moved camera.

Both approaches do not require the measurement of the reference object that is necessary for the calibration procedure. Both approaches use knowledge about the relative movement between the camera and reference object. The robot performs the movement. Image information and the knowledge about the movement are sufficient to obtain the model parameters. Both approaches belong to the category of the active self-calibration approaches. The essential differences can be found in the configuration of the camera and the unmeasured reference object. A camera is rigid and a reference object is connected with a robot in the first approach, whereas in the second approach a camera is connected with a robot and a reference object is rigid. The lens distortion is taken into consideration in both approaches. One parameter for the lens distortion is only used in the first approach, whereas the distortion model is more complex in the second approach. The approaches were evaluated. For both approaches more calibration points are recommended as necessary for the solution systems. The principal point coordinates are calculated for every calibration point. The mean value and the standard deviation are computed over all experiments. To assure the reproduction of the result several calibrations are performed by the same system configuration. The calculated standard deviation is called the R-standard deviation. A high R-standard deviation shows that the calibration approach is inappropriate, because the values of the single experiments differ strongly and reproduction is not possible. Reproduction and deviation are both quality characteristics to evaluate the system, whereas reproduction is the more important criterion for the evaluation, because a poor reproduction is an indication of an inexact parameter determination. The number of reference points is varied to see the dependency between the results and the number of reference points. The reference points are arranged on the object in snake form, see Figure 60.

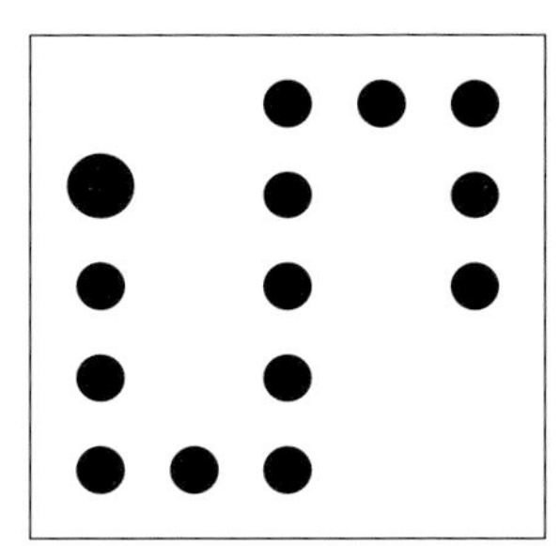

**Figure 60** Reference object with points in snake form [33]

Two steps are executed:

1. The start point is determined.
2. A sorting occurs with regard to the criterion of the shortest distance.

Figure 60 shows that one point is larger than all the other points on the object. This point is the start point and can be found in the first place of the result vector. A function for the sorting is utilized that uses the criterion of the shortest distance. The point that is the nearest to the start point, is chosen. This point is then the second point in the result vector and serves as the new start point.

### 7.2.1
### Calibration with Moving Object

The approach of Lin and Lauber [111] does not need a precisely measured reference object in contrast to conventional calibration approaches. The reference object is connected with the robot in the approach. A camera then takes images from the robot in different positions. A 'quasimeasurement' is obtained with the movement of the robot. The quality of the measurement is determined from the precision of the robot. The positions are taken sequentially by performing different kinds of movements, like translation and rotation alone or together to change from one position into the next position in the sequence. Figure 61 shows such a sequence.

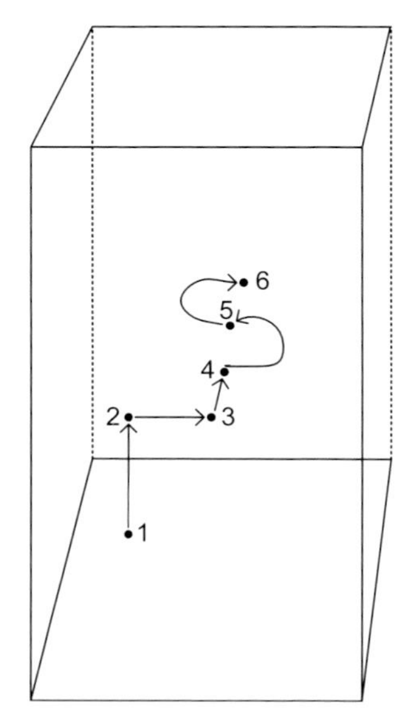


**Figure 61** Six positions of the robot [33]

The first image (1) is taken. Then the robot executes three translations (2, 3 and 4). The positions 5 and 6 are reached by using rotation and translation. The approach is separated into two parts, whereby part 1 consists of two steps:

Part 1
Step 1: The internal parameters and the camera orientation are calculated here.
Step 2: The calculation of the camera position is accomplished.

Part 2
The calculation of the lens-distortion coefficient and the iterative improvement of all internal parameters is the task of part 2.

The lens-distortion coefficient $\psi$ is set to zero in the first iteration step of part one. The lens-distortion coefficient and the other parameters are improved iteratively in the second step. As mentioned, a pair of points is needed to calculate the parameters. This means that two points on the reference object at least must exist. The approach was tested with a zoom camera but the results revealed that it was not appropriate for the calibration of a zoom camera.

### 7.2.2 Calibration with Moving Camera

The approach of Wei *et al.* [112] is considered here according to the explanations in [33]. The camera must be connected with the robot if the approach is to be applied. The reference object is rigid. The robot moves to different positions and takes an image of the reference object from every position. The reference object must contain at least one point. Then the projection parameters and the camera coordinates are calculated by the position alteration(s) of the point(s) in an image and the known robot movements. The model for the lens distortion is more complex here in comparison to the approach of Lin and Lauber. The radial distortion and the tangential distortion are modeled with two parameters. The defined movements of the approach are shown in Figure 62 [33].

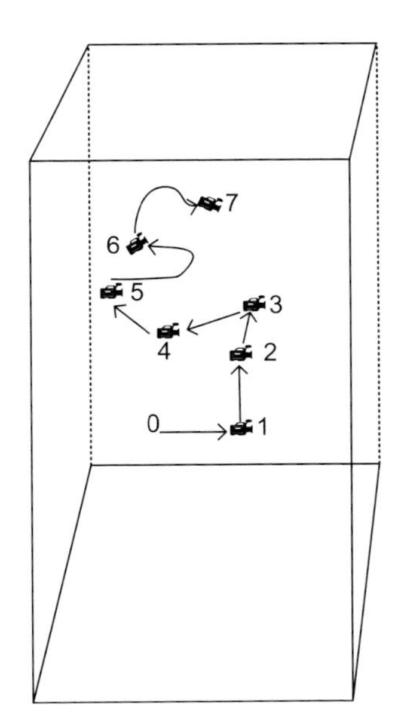


**Figure 62** Seven positions of the robot's camera [33]

First, the robot takes the position zero. Five translation movements can be observed in the sequence. These are the positions one to five in Figure 62. Three positions must be linearly independent to get a solution of the equation system. Then two additional positions at least are necessary. Both must be gained by using translation and rotation together to change from one position into another position of the sequence. It is permitted that the robot can take in further positions to improve the calibration results. The approach is analogously structured to the approach of Lin and Lauber. It also consists of two parts. The first part is decomposed into two steps:

Part 1
Step 1: The movements based on translation are used to obtain the initial values of the internal parameters: the three angles of the camera orientation and the coordinates of the reference points used.
Step 2: The movements, which consist of translation and rotation together, are used to calculate external parameters that are robot coordinates representing the camera position or tool coordinates when a gripping device is mounted onto the robot.

Part 2
In this part, parameters are calculated iteratively that characterize the lens distortion.

Tests have shown better results in comparison to the approach of Lin and Lauber. But the reproduction of the results was as poor as by Lin and Lauber because of the high deviation.

# 8
# Self-learning Algorithms

Neural networks are based on the human brain. Therefore, a brief introduction into the human brain is given here. The human brain consists of neurons. Inhibiting and exciting connections exist between the neurons. A weight is assigned to each neuron. The stimulation and blocking of a neuron is responsible for the value of the neuron weight. So a set of a certain input is represented in certain regions of the brain. The brain is self-organizing, because similar inputs are projected into areas of the brain that are at close quarters. Complex tasks like the control of the fine motor manipulations need more area than rather simple tasks [49].

The self-organizing maps (SOMs), which were developed from Kohonen, are based on the human brain. The SOM consists of $n$ neurons, which belong to the set $SE$. A weight vector $\nu_i \in R^m$ with elements $\omega_{il}$, $l = 1, 2, \ldots, m$, is attached to every neuron $v_i$, $i = 1, 2, \ldots, n$. The distance between two neurons $v_i$ and $v_j$ in the map can be measured with $\mathrm{d}_{SE}(v_i, v_j)$. Vector $ip \in R^m$ represents an input (training) signal. Similarity between the input signal $ip$ and the weight $\omega_{il}$ can be determined with the Euclidean distance $||.||$. A neuron $v_{i^*}$ is activated from the input signal $ip$ if the evaluation of inequality provides 'true' as a result [113]:

$$||\nu_{i^*} - ip\,|| \;\leq\; ||\nu_i - ip\,||, \;\; \forall\, v_i \in SE. \tag{8.1}$$

The neuron with the smallest Euclidean distance between its weight vector and the input signal fulfils the inequality. Working with neural networks involves generally two parts. In the first part the neural network has to be trained with input data, which are similar to the input at runtime. In the second part the trained neural network is used at runtime to recognize the input. The training of SOMs comprises three steps. The initialization takes place in the first step. So the weight $\omega_{il}$ of every neuron $v_i$ gets a value assigned that can be calculated randomly. In the second step an input vector from the training set is chosen in response to obtain an activated neuron, which is called the 'winning neuron' $v_{i^*}$. The selection of the input vector follows a probability distribution. Then an adaptation occurs in the third step. The weights $\omega_{il}$ ($v_i \in SE$) of the neurons in the $SE$ neighborhood of $v_{i^*}$, which can be determined with the distance $\mathrm{d}_{SE}$, are modified. The steps two and three can be repeated several times. The calculation of the weight $\omega_{il}$ of a neuron $v_i$ at iteration step $t$ occurs according to the following formula [113]:

*Robot Vision: Video-based Indoor Exploration with Autonomous and Mobile Robots.* Stefan Florczyk

ISBN: 3-527-40544-5

$$\Delta\omega_{il}(t) := \mu(t) \cdot \phi_t(v_{i^*}, v_j)(ip(t) - \omega_{il}(t)), \tag{8.2}$$

| | |
|---|---|
| $ip(t)$ | Training input at iteration step $t$ |
| $v_{i*} = v_{i*}(ip(t))$ | Winning neuron at iteration $t$ |
| $\mu(t)$ | Learning rate at iteration $t$, which is a monotonic decreasing function with $0 < \mu(t) < 1$. |
| $\phi_t$ | Activation profile at iteration $t$. The closer the weight $\omega_{il}$ of the neuron $v_i$ to the weight $\omega_{i^*l}$ of the neuron $v_{i*}$ the more it is activated from the neuron $v_{i^*}$. |

The new weights of the neurons are assigned with $\omega_{il}(t+1) := \omega_{il}(t) + \Delta\omega_{il}(t)$ after the computation of $\Delta\omega_{il}(t)$ has taken place. So the neurons within a neighborhood have similar weights. This means that similar data can be found in such a cluster. So the neural network is used to recognize data at runtime in the second part. It is necessary to count the activations of the neurons. The recognition of the input data is performed by the use of the neuron's activations [113].

## 8.1 Semantic Maps

Semantic nets can be derived from semantic maps. Semantic nets are graphs that have nodes representing terms and edges representing associations. It is possible to cluster the attributes of objects in a semantic map with the aid of SOMs. The clusters can then be converted into a semantic net. An input set of 16 animals was used

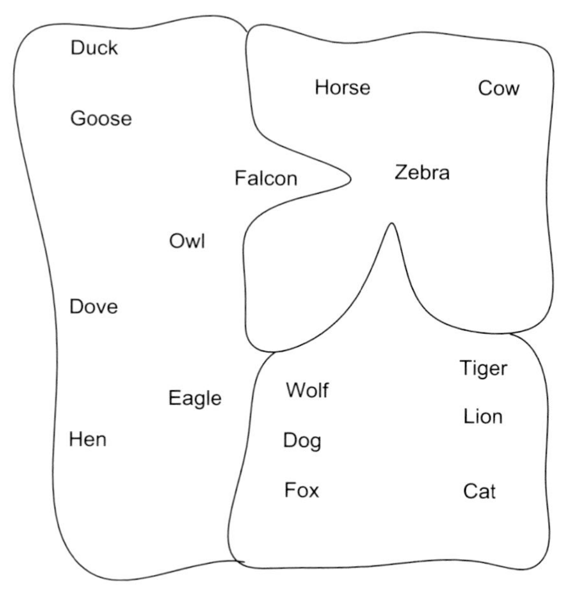


**Figure 63** Semantic map [49]

in an example. Each animal was characterized with a 13-dimensional attribute vector. The attribute vectors were used to train a SOM in 2000 learning steps. The weights were chosen initially by random. An animal is attached to the neuron that has the highest activation for the particular animal. Three main categories can be found in the map shown in Figure 63 after the training [49].

Hoofed animals are positioned in the top-right part, birds in the left part and wild animals in the lower-right part of the map. The distance between the attribute vectors of neurons, which are located in the same part, is less than the distance between the attribute vectors of neurons, which can be found in different parts [49].

## 8.2 Classificators for Self-organizing Neural Networks

Self-organizing neural networks can not be used for classification. An additional level must be built, which is responsible for the classification and obtains the input of the self-organizing neural network for interpretation. This requires a training method, which consists of two steps [49], see Figure 64.

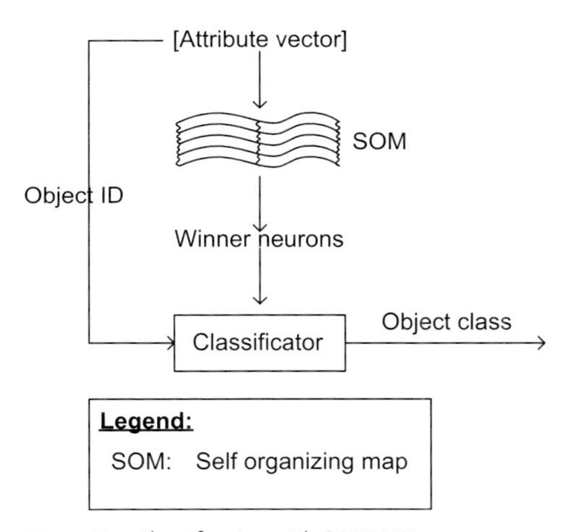


**Figure 64** Classification with SOM [49]

The training of the SOM occurs in the first step. After the training is finished, the coordinates of those winner neurons, which represent a certain object identity, serve as input patterns for the training of the classificator. This can be a linear decision maker, which selects the winning neuron that has the highest activation. The linear decision maker is a supervised learning classificator in conjunction with the technique of self-organization, which is supported by the SOM. This means that the classificator must know to which object class every output neuron belongs. The classificator is not able to generate clusters for similar inputs, which is a feature of the self-organization. It is only possible to identify the class affiliation of a neuron if it was activated during the training phase. This is a drawback of the classificator, because the class affiliation is only known for some of the neurons. These considerations result in some requirements that a classificator should have if it cooperates with a SOM [49]:

1. The classificator should use the feature of the cluster creation provided by the SOM. Linear decision makers are only able to connect neurons that have been activated during the training phase, to the object class. Hence, only a part of the SOM's output layer is used as input for the classificator, but the entire output layer should be used from a decision maker.
2. The classificator should be self-organizing as well as the SOM. The classificator should find the cluster boundaries without help only by analyzing the neighborhood of those neurons that have been activated during the training phase.
3. The misclassification of a neuron to a cluster must be avoided by using the affiliation criterion. For these purposes the distance from the examined neuron to the center of the cluster and the weight vector of the neuron can be utilized, which depicts a prototype of an object.
4. Ambiguities should be adopted, because different three-dimensional objects can have the same views, which are represented by the same neuron in the output layer of the SOM. This results in overlapping but correct clusters. These overlapping clusters should be reflected in the classificator.

The approach of adaptive cluster growing (ACG) uses a classificator for SOMs that supports the cluster recognition in the SOM. The ACG uses neurons that have been activated during the training phase. At the beginning it is known only for these neurons to which object class and cluster, respectively, they belong. The ACG is a one-layer neural network whose input neurons are the output neurons of the SOM, whereby every output neuron of the SOM is connected with every output neuron of the ACG. So a many-to-many connection exists here. The procedure starts, for example, with an input image $I$ for the SOM that examines the winning neuron. The connections between the SOM neurons and the ACG neurons have the value null if the winning neuron is not activated from an object class that is represented by the connected ACG neuron and one if the winning neuron is activated by the object class. It is possible to model ambiguities with this architecture, because one winning neuron can activate a part of the entire neuron set [49], see Figure 65.

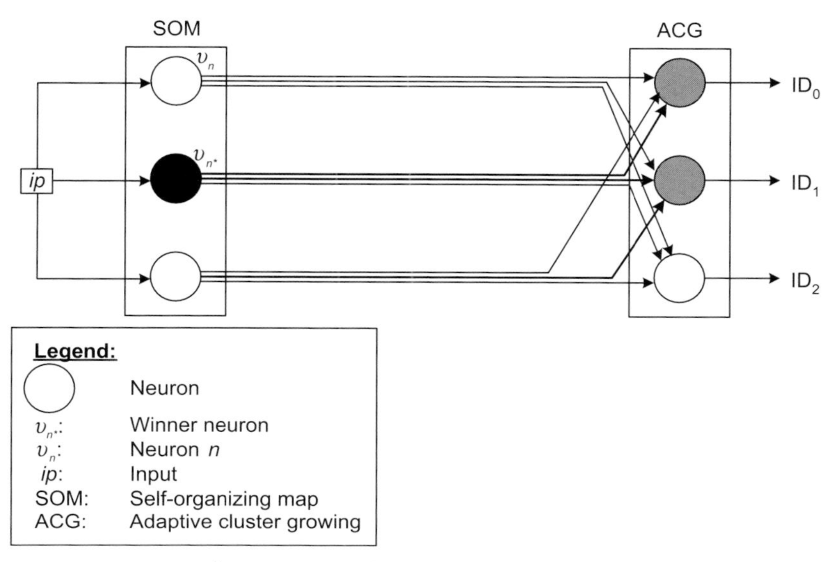


**Figure 65** Connection between SOM and ACG [49]

Connections between the SOM and the ACG with value null are represented by thin lines and connections with value one by thick lines. $\upsilon_*^{\text{SOM}}$ is the winning neuron and activates the two neurons $\upsilon_0^{\text{ACG}}$ and $\upsilon_1^{\text{ACG}}$ in the ACG, because the value of the connection is one. The connection to the neuron $\upsilon_2^{\text{ACG}}$ in the ACG has the value null. Therefore, this neuron is not activated from $\upsilon_*^{\text{SOM}}$. The neuron $\upsilon_0^{\text{SOM}}$ in the SOM activates no neurons in ACG. This can happen if the input for the training of the SOM did not activate this neuron, or during the training of the ACG it was not possible to relate $\upsilon_0^{\text{SOM}}$ to an object class and cluster of the SOM, respectively. Therefore, a further training is required to detect the clusters. This is a recursive algorithm that analyses the environment of the activated neurons in the SOM. Whether the activated neuron and a further neuron are similar can be decided by the comparison of the two prototype vectors. The similarity between the two vectors can be proved with a threshold. The threshold is increased in every recursion step. Hence, it can be assured that neurons that are further away from the center must have a higher similarity, and the threshold serves as the break criterion for the recursive algorithm. This is shown in Figure 66 [49].

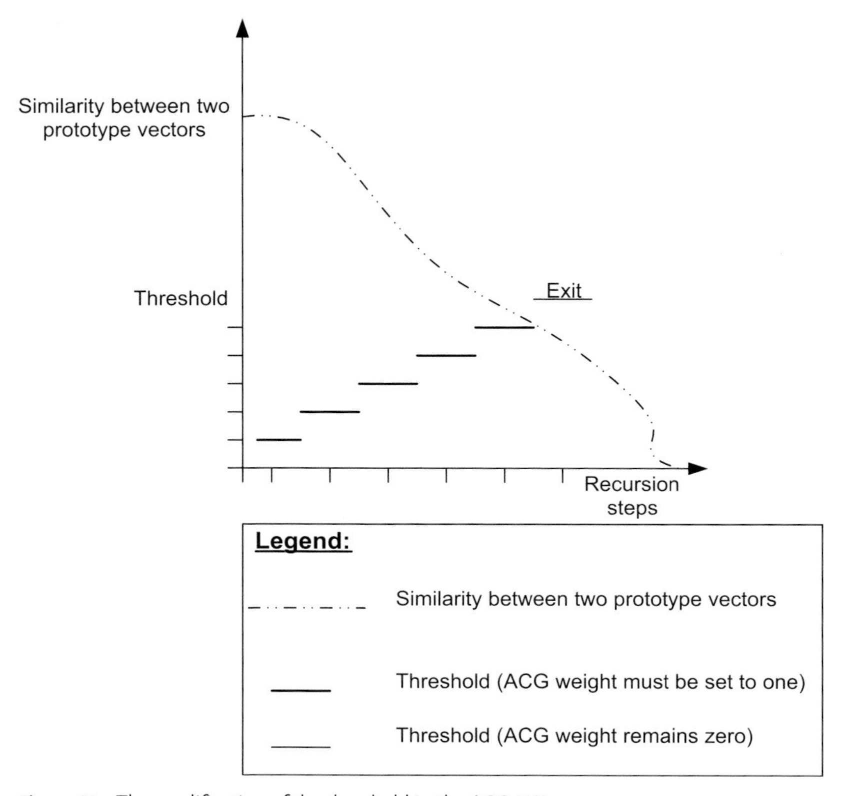


**Figure 66** The modification of the threshold in the ACG [49]

The recursion steps are written at the $X$-axis. The interrupted line represents the similarity between two prototype vectors. A horizontal line can be found in every recursion step. The line represents the threshold that is increased in every recursion step. If the horizontal line representing the threshold is painted large, then the weight of the connection between two neurons in the ACG must be set to one. If the line is painted thin, the weight remains zero. This emerges in recursion step six, because the threshold exceeds the similarity between two prototype vectors [49].

# 9
# OCR

Commercial OCR systems are able to recognize characters that have a size of at least 6 until 8 points character height. It is not possible to increase the recognition rate with a size alteration of the characters. To get better recognition results, it can be a good strategy to perform manipulations at the characters. This can be executed by using operators that are known from the image processing (see Chapter 2) like morphological operators, skeleton operators, and so forth. Character recognition is often difficult because the characters can be stuck together. It must be said that it is not always possible to recognize characters in a correct way, when they are stuck together, because an interpretation of such characters can be achieved differently. This is shown in Figure 67 [114].

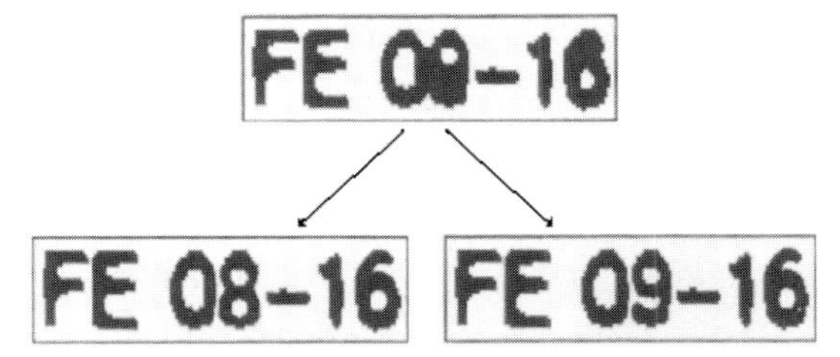


**Figure 67** Ambiguity in character recognition [114]

Commercial OCR systems have been tested. The tests have shown that the increase of the resolution from 300 dpi to 600 dpi has not provided an essentially better quality in character recognition [114].

Further errors in character recognition can occur. Characters that are close together in an image are often stuck together, see Figure 68.

*Robot Vision: Video-based Indoor Exploration with Autonomous and Mobile Robots.* Stefan Florczyk

ISBN: 3-527-40544-5

**Figure 68** Characters that are stuck together

Figure 68 shows the room number of an office on a doorplate. Two numerals are stuck together. This fact makes a correct recognition difficult. But the merging of two different characters is not the only problem in OCR. Sometimes a merging can occur within the character, see Figure 69.

**Figure 69** Merging within a character

The doorplate in Figure 69 now contains a room number that has a numeral stuck together in itself. It is also a problem in OCR when a character has a similarity to another character, where it can be interpreted as that character to which the analogy exists, as in Figure 70.

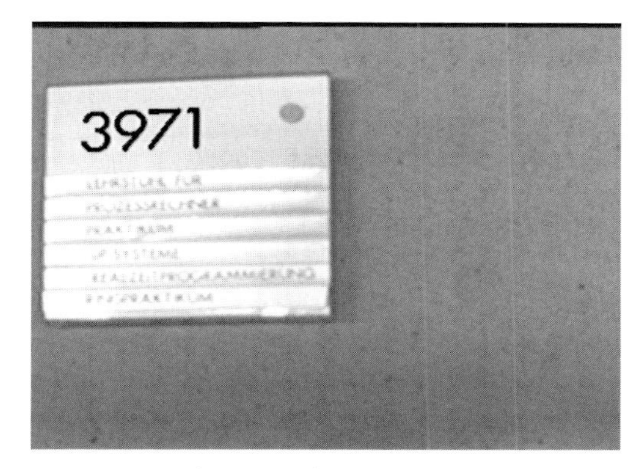


**Figure 70** Similar numerals

The two numerals one and seven in Figure 70 are part of the room number in the doorplate. Depending on the quality of the image it can happen that the one is interpreted as the seven and vice versa. The recognition process is sometimes difficult if some characters are not completely closed, as in Figure 71.

**Figure 71** A numeral that is not closed

The room number on the doorplate in Figure 71 has a numeral that is not closed. This is the six. Therefore, the recognition of the six is problematical.

# 10
# Redundancy in Robot-vision Scenarios

A redundant robot-vision program is explained in this chapter, which was developed for a robot-vision scenario. The explanations are based on [115].

A mobile robot autonomously creates a map (see Chapter 3) of its office environment in this scenario (see Chapter 1). After the map was created, it serves as the basis for the robot to fulfill tasks executed by an office messenger or a watchman. For example, postal delivery could be a task for an office messenger. The robot can only fulfill such tasks if it possesses cameras. The robot must be able to read door numbers that can be found on office doorplates. Reading the office numbers is not always simple, because the dynamic character of the elucidated scenario aggravates the recognition process. The working time of the robot is not restricted generally. This effects inhomogeneous illumination. The robot will take images at different times of day. Therefore, a robot-vision program that can read the room numbers must consider different natural and artificial illumination intensities.

Another problem are different positions from which the robot takes images from the doorplate. This may concern the angle that is determined by the camera's line of sight that meets the doorplate. As the angle becomes more acute the more difficult it is for a robot-vision program to read the numbers on the doorplate. Images that are taken from an acute angle can affect problems that are well known in OCR (see Chapter 9). These problems can yield poor recognition of the numbers. It is not possible to handle the problems with commercial OCR programs. Therefore, the implemented program uses the commercial image-processing library HALCON [4]. HALCON provides a self-learning algorithm that can be trained with digits. Also, neural networks (see Chapter 8) have been successfully used for OCR. Therefore, several approaches exist that can be used for OCR in computer-vision programs. But an application must nevertheless be well designed for the specific problem to gain good recognition results. This requires proper image segmentation.

The segmentation process can be supported by several operators. Frequency filters, such as a highpass filter, a Gabor filter, etc., can be qualified for the segmentation of images that were taken under different illumination conditions (see Section 2.2). So there is probably no need for the development of new operators. A redundant program design is proposed in this chapter. It is shown how redundant robot-vision programs are designed to obtain good recognition results. The implemented program processes several results that are compared to select a solution with the

*Robot Vision: Video-based Indoor Exploration with Autonomous and Mobile Robots.* Stefan Florczyk

ISBN: 3-527-40544-5

highest recognition rate. If the digit recognition was not satisfactory, the program processes additional results to obtain a better solution.

An overview is now given of the program design. General guidelines are also introduced for the development of redundant programs in computer vision, which are derived from the program that is explained in the following. For these purposes the program flow is considered that manifests the redundant calculation. The program was tested in an experiment. Results are reported. The used sample images show doorplates and have been taken in different conditions, like changes in illumination and the position from which the images were taken.

## 10.1 Redundant Programs for Robot-vision Applications

It has been discussed that the generation of a map, which will be the base for the fulfillment of tasks that an office messenger or a watchman has to do, with an autonomous and mobile robot is very difficult and can only be solved with the support of a robot-vision program. Because of the dynamic character, it is a challenge to create a robust program. In particular the changes in illumination can quickly yield a failure of the robot-vision program. A redundant program design is proposed to improve the recognition process. The quality of the recognition process can be raised with different strategies.

The segmentation process can be advanced. It may also be a useful approach to construct a reliable classification. Classification is the last step in computer vision (see Chapter 2). In this step attributes of objects are examined and compared with a class description. If the examined values match the class description, the objects are assigned to the class. For example, a segmented rectangle in an image can represent a doorplate in an office environment. The doorplate's measurements, like the width and the height, can be attributes for the classification process. The aim of the developed program was the appropriation of a robust segmentation for the classification. Several approaches, like the elucidated neural networks and self-learning approaches, exist for OCR. It seems that there is no need for the additional development of classification techniques. Rather a good segmentation process is required that provides qualified results for the classification. The classification can only process useful results if the segmentation process is successful.

In this section it is not proposed to develop new operators. The main focus was the development of a redundant program design to meet the dynamic character of the robot-vision scenario. The program uses two redundantly calculated results, which are used for the examination of the final result. This procedure realized the principle that a final solution must be based at least on two redundantly calculated results. The consideration of a single result is regarded as too insecure. It is also required that calculations be carried out with different ranges. For example, the distance between a robot and a doorplate can vary in the elucidated scenario in certain boundaries. Although the distances do not differ strongly, the fluctuations can affect the segmentation. For example, if the segmentation process is searching for a rectangle, which probably represents a doorplate, it can use the rectangle's measurements in the image for the detection. But these measurements will vary if the distance changes between the robot and the doorplate.

Therefore, it is proposed to use redundant calculations with different parameterizations. But a different parameterization is not sufficient. The redundant calculations will also be accomplished using different strategies. This concerns the implemented algorithms. In the elucidated scenario it could be a way first to search for a rectangle that may represent a doorplate. If the rectangle is found, the further calculations can be restricted to the detected rectangle, because the room number is expected to be within the rectangle. But it can also be successful to try a direct recognition of the room number in the entire image. In this case the rectangle search is circumvented, which can be especially valuable if the detection of a rectangle is difficult. For example, an image that contains only a part of the whole doorplate would aggravate the rectangle detection. If it is not possible to find a rectangle, it is necessary to generate both redundant results with direct character recognition. These explanations reveal that a redundant program can demand a very high programming complexity. It depends on the application whether this can be justified.

Also, a long running time can result from redundant programming that should be palliated with another design principle. First calculation attempts should always use fast algorithms. Therefore, it is approved that these calculations must not be very precise. More precise algorithms are only utilized if the first calculations are unsuccessful. In this case a longer running time is expected. Also, plausibility checks can be used to improve the quality of the results. For example, it can be presumed that office room numbers consist of more than one digit. If a final solution contains only one digit, it is recommended that further calculations be executed. The examined final solution is obviously incomplete. The explained design principles are now listed:

1. A final solution must be based on at least two calculations.
2. Different strategies should be used for the calculations.
3. Redundant results should be gained using different parameterizations in the recognition process.
4. Fast calculations are required for the first calculation tries.

The redundant program follows these guidelines and is explained in the next section.

## 10.2
## The Program

The functionality of the program functions and the program flow is elucidated. To meet the guidelines the functions often offer a relatively large number of formal parameters to support the use of different parameterizations. Images are analyzed with operators that are parameterized very loosely with wide parameters. This means that the first calculations possibly select many details in the image in order to avoid necessary information being eliminated. Otherwise, if too many details are selected, the character recognition can also be aggravated. For example, wide ranges can be provided for the feature height. A rectangle can be searched for in an image with a relatively wide interval for the feature height to avoid the possibility that the rectangle that represents the doorplate will not be selected.

But it may also be the case that rectangles are found that do not represent the doorplate. In this case the unwanted objects must be eliminated in a downstream processing level. The program implements two different strategies for the character recognition. It performs a direct recognition of the characters in the entire image, or it first searches for a rectangle that probably represents a doorplate. Both strategies are explained in the following.

### 10.2.1 Looking for a Rectangle

The strategy that first looks for a rectangle is implemented with the function `rectangle_search`:

```
void rectangle_search (int MEAN_VALUE_SEARCH, int
Height_MIN).
```

First, the function is searching for edges with the Sobel operator (see Section 2.4) and processes edge data like lengths, coordinates, and orientations. But it must be noted that the edge detection can be very time consuming. This depends strongly on the image in which the edges are to be detected. For example, documents with text and many figures can often be found near to an office door. In this case a vast quantity of edges will be processed that can take a long time. A mean value filter is applied to the image to get a smoothed version and thereafter the Sobel operator is applied. This procedure will help to save processing time, because the smoothed image will contain fewer edges. The user of the function `rectangle_search` can control if the mean value filter is taken or not. Therefore, the flag `MEAN_VALUE_SEARCH` is offered. This flag must be initialized with one if the mean value filter is used before the edge detection.

With the second parameter, the minimal height for objects that are to be selected can be adjusted. Objects whose region heights fall below this lower limit are not considered for further calculations. The calculated edges are approximated with lines [4] by the use of a HALCON operator. This creates a binary image. Also, the mentioned edge data are used for the processing of these lines. Then the lines are elongated. The extension of the lines generates a rectangular shape that has crosses at its corners, because in many cases the segmented rectangle, which represents the doorplate, has interrupted lines.

It is now expected that the rectangle is bordered by a closed contour. Holes are now filled. This results in a filled rectangle. The binary image is processed with erosion and dilation (see Section 2.3). Until now the binary image contains exactly one region. This region is decomposed. Coherent pixels in the image constitute a region (see Section 2.6). The expected size features of the rectangle, which probably represents the doorplate, are now used to eliminate the unwanted regions. Only the rectangle remains. As elucidated, the possibility exists that the function `rectangle_search` can be parameterized in such a way that the edge detection is executed on a smoothed image. This saves processing time and should be used for the first calculation tries. If these attempts do not provide the required rectangle the procedure can be used once again. But now the smoothing is omitted. A more detailed and more time-consuming search is then performed. The recognition of the room number can start if the rectangle is found. The function `rectangle_search` uses a variable to indicate if the rectangle detection was successful.

### 10.2.2
### Room-number Recognition

Room-number recognition can start if the rectangle is found. Therefore, the function `char_rec` can be used:

```
char_rec(int NO_REC, int RawSeg, int Height_MIN, int
Height_MAX, tuple RecNum_REC, tuple Confidence_REC).
```

At first the parameters are depicted. `NO_REC` is a flag that indicates if a rectangle was found or not. Zero indicates that a rectangle was found. The function uses the dynamic threshold (see Section 2.7) of the HALCON library. Objects that are selected by the function must have a region height of at least `Height_MIN` pixels and may have the maximal region height of `Height_MAX` pixels. The recognized digits are stored in `RecNum_REC`. For every recognized digit a confidence value in per cent is calculated. This value indicates how secure the recognition of the corresponding digit was.

At first the function starts with the cutting of the calculated rectangle. The result is an image clip that is augmented. A homogeneous matrix (see Section 3.1) is used for these purposes to execute translation and scaling. Then the image clip is processed with a mean value filter, which follows the dynamic threshold. A similar effect could be gained if a threshold operator were applied to a highpass filtered image. The dynamic threshold operator offers a parameter to extract light or dark areas. Room numbers are often in black. In this case the parameter must be set to 'dark'. Holes, which may result from the application of the dynamic threshold operator, are filled if they match to a specified size. An opening is now used that consists of erosion followed by the Minkowski addition to calculate smoothed rims. Coherent pixels are now analyzed that constitute a region. Regions can now be selected that match to the expected minimal and maximal height and width of digits.

At best only those regions remain that represent the digits of the doorplate. The recognition of the digits can then start. For these purposes a self-learning algorithm [4] of the HALCON library is used that was trained with eligible digit examples. The result of this training is an ASCII file that is used at run time from an operator that also determines the confidence values. The confidence values can be used to eliminate regions that do not represent digits, because the corresponding confidence values should be very low in this case, unless the region has a strong similarity with a real digit. But this would be random, and it can be expected that this should not emerge too often. It may be more frequent that regions that do not represent digits and do not appear similar, have measurements that match to the expected height and width of digits on a doorplate. These regions can be eliminated by the use of the confidence values. The program uses a boundary for confidence values. Only regions that have a corresponding confidence value that does not fall below the boundary are selected into the result. In any event, a region that represents a digit should never be eliminated. Therefore, the boundary was chosen very low so as to avoid this case, but it may now occur that unwanted regions are sometimes selected.

The redundant program design helps to detect such erroneous classifications. The obtained result can be compared with a second redundantly calculated result. This result is examined with the direct recognition of digits in the entire image.

### 10.2.3 Direct Recognition of Digits

The operator for OCR can also be used from an algorithm that implements the direct recognition of room numbers, because the size of the selected digits has no effect on the classification. The function `direct_recognition` implements the algorithm:

```
direct_recognition(int Width_MIN, int Width_MAX, int
Height_MIN, int Height_MAX, tuple RecNum_TH, tuple
confidence_TH).
```

The user of the function can provide minimal and maximal boundaries that specify allowable intervals for the height and width of regions that may be selected. The recognized numbers are stored in the array `RecNum_TH`. The corresponding confidence values can be found in the array `confidence_TH`. The function starts with the mean value filter. It follows the dynamic threshold. Specified holes are filled in the binary image. Opening with a circle follows this. The entire region is decomposed into regions that represent coherent pixels as elucidated before. The values provided for the allowable boundaries of object width and height are used to select regions. The boundaries must be lower than those boundaries that are used in an algorithm that first searches for a rectangle, because the digits are smaller. In particular no augmentation happens in the algorithm that implements the direct recognition.

As soon as the regions that probably represent digits are detected, the classification can start, which is similar to in the function `char_rec`. An operator that uses an ASCII file that was calculated from a self-learning algorithm, performs the digit recognition. Arrays are also used to store the recognized digits and the corresponding confidence values. Figure 72 shows the direct recognition of digits in an image.

**Figure 72** Direct recognition of a room number

The number on the doorplate is '4961'. The recognized digits are bordered with rectangles. The digits read are written into the top-left corner of the figure.

### 10.2.4 The Final Decision

Two results are now calculated with different strategies. It is now the task of the program to make a final decision. The best result is selected. At first it is checked if two results exist, because only in that case can a comparison be executed. The flag `NO_REC` serves for these purposes. If `NO_REC` has the value zero, two results exist. If `NO_REC` does not have the value zero, a further result must be processed, because a final decision can only be executed if two redundant results are available. The comparison of two redundant results is performed with the function

```
result_intersection (tuple Result1, tuple Result2, tuple
Intersection).
```

The two results are provided by the use of the formal parameters `Result1` and `Result2`. The function returns the intersection in the variable of the same name, which is also a formal parameter. Only digits are included in the variable `Intersection` whose corresponding confidence values do not fall below a boundary in both results.

## 10.3 The Program Flow

The OCR file that was generated from a self-learning algorithm is first read from the program. The call of the function rectangle_search follows. At the beginning a result is sought by the use of the mean value filter. Therefore, the flag MEAN_VALUE_SEARCH is initialized with one. This affects a fast but not very precise search. It is then checked whether a variable contains only one region that probably represents the desired rectangle. If one region is contained exactly in the variable, the function char_rec is called. If the number that indicates the contained regions is not equal to one, the calculation was not successful. A further call of rectangle_search is required. The flag MEAN_VALUE_SEARCH now obtains the value zero. The mean value filter is not applied in this case. Therefore, the proposed guidelines are considered. If the fast search with the utilization of the mean value filter is not useful, a further search is executed that performs a more precise but also more time-consuming search. Figure 73 shows the entire program flow.

After the function rectangle_search has finished, a variable should contain a rectangle that represents the doorplate. This is verified and if true, the function char_rec is called. Otherwise the detection of the doorplate was not successful and rectangle_search is called once again. MEAN_VALUE_SEARCH is now initialized with zero to indicate that the segmentation should not use the mean value filter. The mean value filter is only applied in the first calculations attempts, because the edge detection in a smoothed image saves running time. But the segmentation process is not so precise. Therefore, the mean value filtering is omitted if the doorplate detection malfunctioned.

A second redundant result is gained with the function direct_recognition that accomplishes the number detection on the entire image. Digits are directly searched for. At first the formal parameters of the function obtain very relaxed values. These relaxed values avoid digits of the doorplate being eliminated. A flag indicates if two results are available and if the rectangle detection was successful. This information can be taken to alter the search strategy if two results do not exist. If a rectangle was not detected, calculation of both results is tried with the direct recognition. A function with the same name is used for these purposes. The values of the formal parameters are now more restrictively selected to avoid unwanted regions being collected. But a restrictive parameterization can have the drawback that sometimes one or several digits are also eliminated.

The function intersection_result obtains two redundant results if available. The intersection of two results is calculated. The quality of the computed intersection is verified with a plausibility check. This check presumes that numbers on office doorplates consist, as a rule, of more than one digit.

If the intersection contains only one digit, the function rectangle_search is called again, but other parameters are used. For example, a former recognition process detected a rectangle that did not represent a doorplate. In this case the parameterization was improper and therefore changed in a fresh calculation. Regions are now also selected if their minimally permissible height is lower than in the former

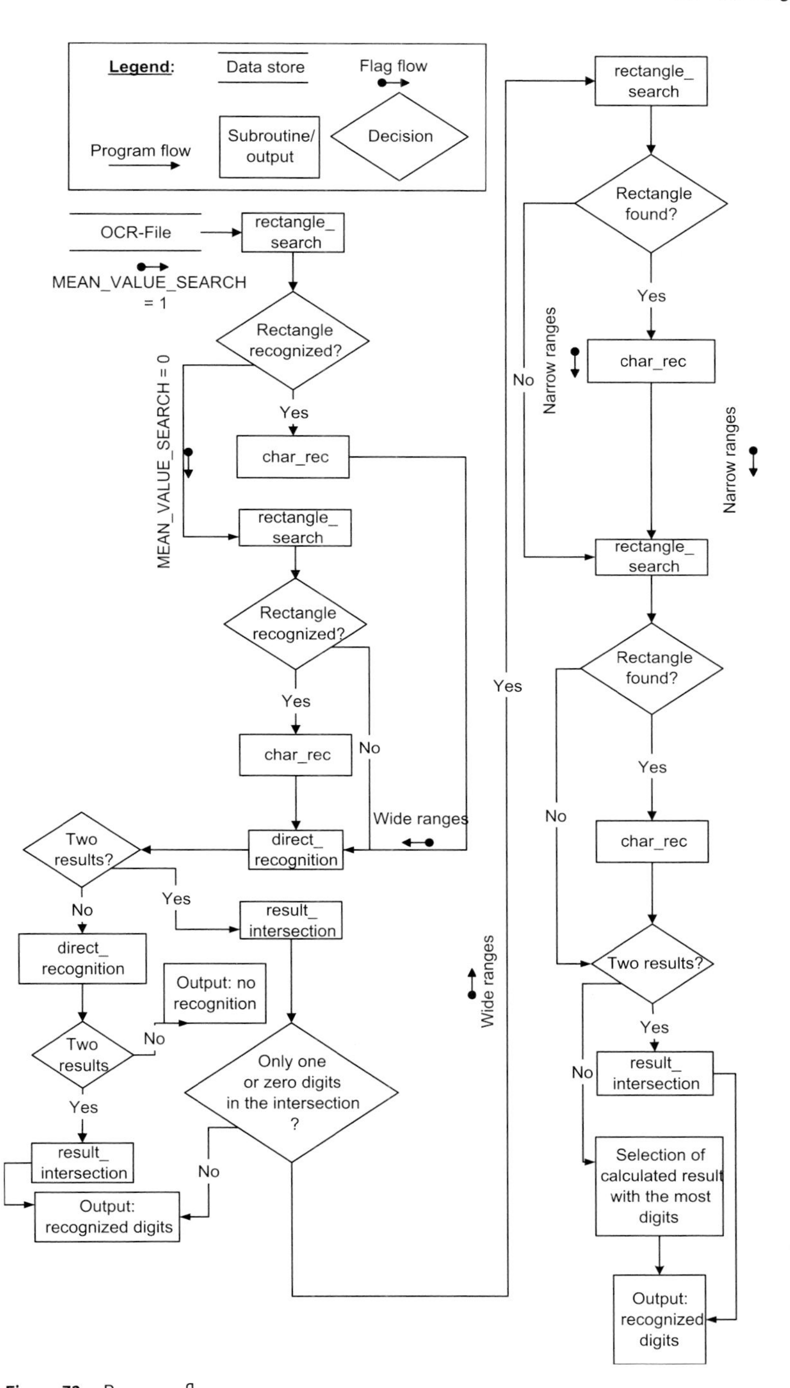


**Figure 73** Program flow

computation, because the value of the corresponding parameter is now decreased. It is verified again if a rectangle is found that is a prerequisite for the function `char_rec`.

If a rectangle is found, the procedure `char_rec` is called with other parameters. The flag `NO_REC_WR` obtains the value one if no rectangle was found. In this case the use of wide ranges (WR) did not also provide a utilizable result. Narrower ranges are now used to calculate a second result with the function `rectangle_search`. If now two new results exist, they were both calculated with the function `rectangle_search`. One result was calculated with narrow ranges and a second result with more relaxed ranges. The function `result_intersection` is now called if both results exist. This is verified.

The final solution is constituted by the intersection from both redundantly calculated results. If two additional results can not be provided, the final solution is composed from the comparison of the former calculated intersection with a probably calculated further single result. It is simply the result chosen that contains the most digits. If no additional result can be created at all, the probably only recognized digit is selected for the final result.

## 10.4 Experiment

Images were taken under different illumination conditions and from different positions. The room numbers were often difficult to recognize also for a human. An example is shown in Figure 74.

**Figure 74** An image of poor quality

Although the room number in Figure 74 is difficult to recognize, it was correctly read from the robot-vision program. An example is also given of an image that was taken from a position where the camera's line of sight met the doorplate in an acute angle, see Figure 75.

**Figure 75** An image with an acute angle to the doorplate

The robot-vision program also read in the room number correctly this case. This holds also for the following two images, which were taken under different illumination conditions. Figure 76 shows an image that was taken in rather dark illumination.

**Figure 76** A dark image

In contrast, a brighter image is shown in Figure 77.

**Figure 77** A bright image

These dynamically changing factors are very problematical for conventional computer-vision programs. The recognition results of the explained program are depicted in Table 1.

**Table 1** The recognition rates of room numbers.

| | Correctly recognized | 3 digits recognized | 2 digits recognized | 1 digit recognized | Digits twisted | Nothing recognized |
|---|---|---|---|---|---|---|
| Quantity | 160 | 14 | 5 | 1 | 4 | 4 |

A total of 188 images were used in the experiment. The room number was read accurately in 160 sample images. Three digits of the entire room number, which consists in all cases of four digits, were recognized in 14 images. Two digits only could be recognized in five images and one digit was recognized in a sole image. In four images the digits were recognized, but the sequence of the digits was erroneously reproduced. No digits at all were recognized in four images.

## 10.5 Conclusion

Robot-vision projects can strongly benefit from a redundant program design. In this chapter a robot-vision program was introduced that can be used from a mobile robot whose task is the autonomous creation of a map that represents the office environment. The generated map can then be a basis for the robot to perform actions that a watchman or an office messenger has to do. For example, postal delivery could be such a task for an office messenger. The reported scenario features a very dynamic character. The robot has to take images under different illumination conditions and from different positions. These factors are a problem for the reliability of the robot-vision application that needs to read the room numbers in the office environment. Several operators and techniques exist for OCR in computer vision. In this chapter a qualified program design was introduced. General guidelines were derived from the redundant program design and reported in this chapter. The use of the general guidelines should enhance the dependability of robot-vision programs in the future. In particular, robot-vision programs that calculate results that are based on probabilities like the self-localization (see Section 3.5), the classification, and so forth, should profit from a redundant program design. The calculated confidence values will be improved with redundant programming. The elucidated program has been in use since 2001 from the autonomous robot system OSCAR [116] for doorplate recognition. OSCAR controls the robot MARVIN. The purpose is autonomous indoor exploration as mentioned before.

# 11
# Algorithm Evaluation of Robot-vision Systems for Autonomous Robots

In the preceding chapter it was stated that redundant programming should be used in robot-vision scenarios. Algorithms are introduced in this chapter that appear to be appropriate for the reported robot-vision scenario. Therefore, the algorithms are compared.

The examples are based on [117]. The evaluation performed was motivated because of the fact that not enough studies exist in the area of the video-based indoor exploration with autonomous mobile robots that report about criteria for the algorithm evaluation. The dynamic character of the robot scenario is considered again. Inhomogeneous illumination can occur. In Section 2.2 it was shown that the Gabor filter can be a good choice if inhomogeneous illumination occurs. The wooden cube could be gripped by the robot also by candlelight. It is not possible to remove the confusions because of the dynamic character of the robot-vision scenario. In this chapter several algorithms are reported that seem to be appropriate for the scenario. The developed algorithms are compared. The comparisons were supported by evaluation criteria that are introduced in this chapter.

Five algorithms based on Gabor filtering, highpass filtering [118], band filtering [4], color-feature detection (see Section 2.1), or Sobel filtering (see Section 2.4) are depicted. Every implemented algorithm uses one of the five listed operators. An image was taken from a Pioneer 1 robot that shows the corridor of our research institute. At first the algorithms are demonstrated, then they are used to detect different objects in the image. For example, a fire extinguisher in an image can be detected with the algorithm based on the color feature. The comparison of the algorithms took place in experiments. The reliability of the algorithms was tested with regard to inhomogeneous illumination. All five algorithms were used to detect the same object in the image. The object is disturbed by a shadow and was therefore selected, because inhomogeneous illumination occurs. The results of the experiment and the used evaluation criteria are elucidated.

*Robot Vision: Video-based Indoor Exploration with Autonomous and Mobile Robots.* Stefan Florczyk

ISBN: 3-527-40544-5

## 11.1 Algorithms for Indoor Exploration

The algorithms are implemented in C++. They are object oriented [119] and use the commercial image-processing library HALCON/ C++. The algorithms are designed similarly. Figure 78 depicts the algorithm that is based on the Gabor filter.

Five classes are shown. The HALCON types `HTuple` and `Hobject` are mainly used. Class `win` is used to show an image `Im` in a window on the screen. The attributes `Height` and `Width` control the size of the window. `WindowID` is the logical number of that window in which the image `Im` is shown. The logical number is assigned from HALCON to every window. If an image `Im` is shown in a window, the user has the possibility to view the image as long as he wants. He can proceed by the execution of an input from the command line. For these purposes the variable `proceed` is used, which stores the user input. Two widow types can be generated with constructor `win(Hobject Image, int standard=0)`. If flag `standard` is not equal to zero, a standard window is generated that has the default settings of HALCON. Otherwise the window size is controlled by the image size. The actual image can be changed with the method `change_image(Hobject Image)`. Image `Im` is displayed with the method `display()`. If the user has the possibility to view the image as long as he wants to, method `hold()` is used. The method ensures that program execution stops until the user performs an input from the command line.

A path to an image that is stored on the bulk memory must be provided as a character array to the constructor `fourier(char file[])` that belongs to the class of the same name. Variable `PioneerImage` is used to store the read image in the main memory. Method `calculate()` performs the Fourier transformation. Variable `ImageFFT` is used for the storage of the transformed image. The class `fourier` can process RGB images. The method `calculate()` in the class `fourier` extracts the blue channel of the RGB image into private member `Blue`. `Blue` is zoomed to quadratic size. The zoomed image is stored in `ImageZoom` and written to the bulk memory. A quadratic image size is often used [110] for implementations of the fast Fourier transform.

Class `gabor` generates a filter of the same name with the method `calculate()`. Constructor `convolution(char file[])` produces a path of an image and convolutes this image with the Gabor filter. The convolution is executed with the method `calculate()` that is called by the constructor. But before `calculate()` is called, the two constructors of the base classes are called. For these purposes the path of the image that is to be convoluted is forwarded to the constructor of the class `fourier`. After the enforcement of the convolution the resulting image is at first in the frequency domain and therefore transformed back into the spatial domain. The image in the spatial domain is stored on the mass storage.

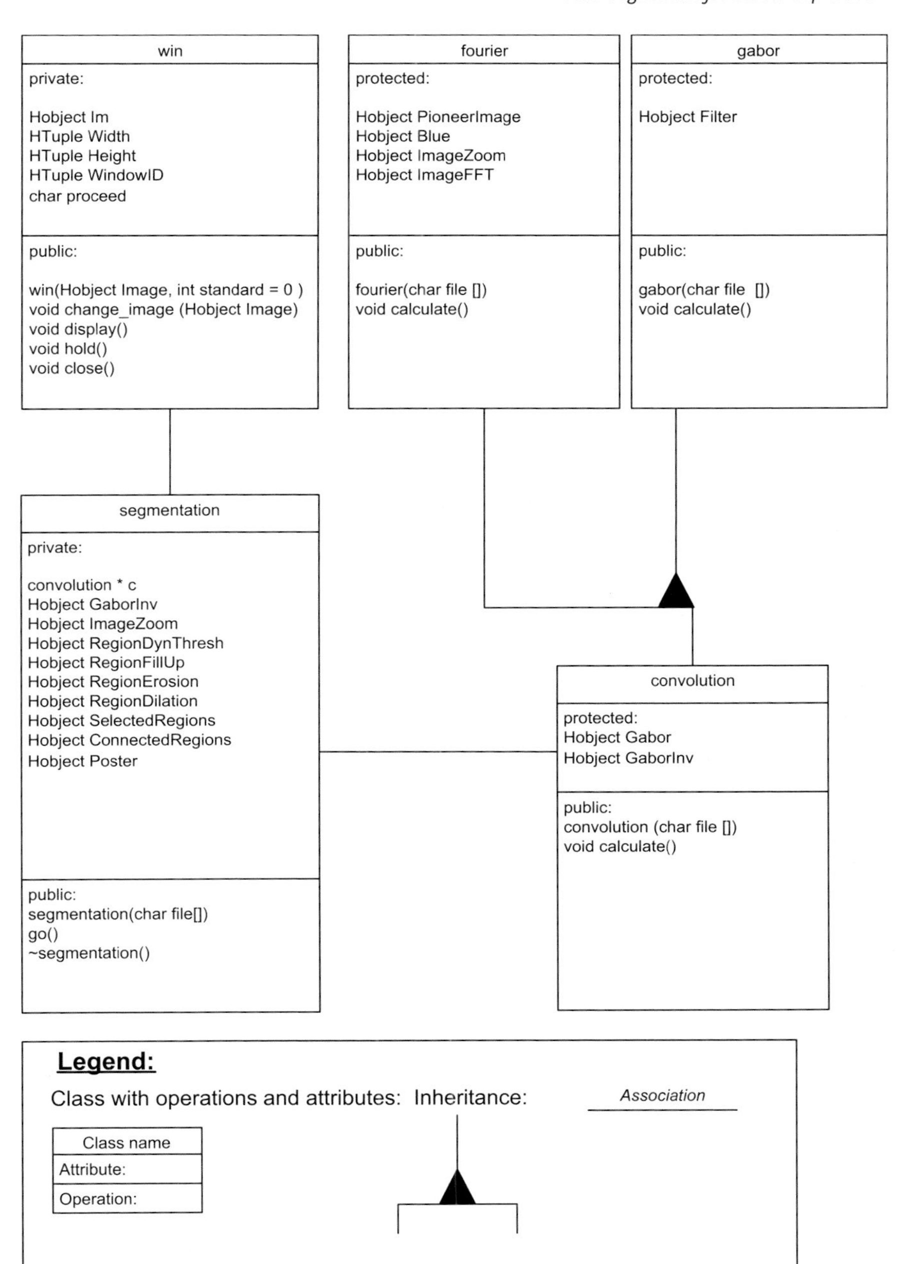


**Figure 78** Class design of an object segmentation algorithm

The constructor of class segmentation creates an object of type convolution on the heap (dynamic store). The constructor of the class `convolution` is called with a path to the image that is to be convoluted. After the convolution takes place method `go()` in class `segmentation` is called. Intermediate data are the convoluted image that was retransformed into the spatial domain, and the zoomed image. Both images are stored on the bulk memory from the base objects as mentioned before. These two results are used from the method `go()`. Extensive explanations of the segmentation algorithm are given below. The segmentation algorithm creates intermediate data that are displayed from an object of the class `win` on the screen. After the segmentation has finished, the desired object remains. Destructor `~segmentation()` is called if an object of the class `segmentation` is removed. The destructor deletes the convolution object from the dynamic memory.

### 11.1.1 Segmentation with a Gabor Filter

Image $I_O$, which was taken from a Pioneer 1 robot, shows a poster that is extracted from the entire image with the Gabor filter. The detection of the poster is difficult because of a shadow cast. This can be viewed in Figure 79.

**Figure 79** Image $I_O$ from a corridor

Two posters can be found on the right side of the wall. The poster nearby the light switch is to be detected. Inhomogeneous illumination can be observed on the poster and its environment due to the shadow cast. Additionally, reflections on the poster result from a neon lamp. Therefore, the poster shows typical emergings that appear in the autonomous indoor exploration with mobile robots.

The Gabor filter was used for the poster segmentation because of its direction filtering effect. The impulse answer of the two-dimensional Gabor filter is represented with the formula (2.26). A total of 50 pixels have been used for the bandwidth $\lambda$. The value of 1 was actually used for $\sigma$ and 0.3 for $\delta$. The Gabor filter was applied to the

Fourier-transformed image that was gained from the extracted and zoomed blue channel of image $I_O$. The zoomed image was of size $512 \times 512$ and denoted with $I_Z$. The result of the Gabor filtering was retransformed into the spatial domain and is denoted by $I_G$ and shown in Figure 80.

**Figure 80** Gabor filtered image $I_G$

The illumination differences of the poster are alleviated. For the image $I_G$ the threshold image was constructed by the use of the dynamic threshold operator (see Section 2.7). The value of 15 was assigned to the offset $of$. Pixels are included in the binary image $I_T$ provided that they fulfill the inequality $I_Z(x_A, y_A) \leq I_G(x_A, y_A) + of$ of Equation (2.46). Morphological operators were applied to the threshold image. The height $H(A_i)$ and width $W(A_i)$ of the desired poster $A_i \in I_E$, $i = 1, 2, \ldots, N$, are used to select the poster. $I_E$ is an image and $A_i$ a region in the image $I_E$. $A_i$ is constituted by a set of coherent pixels. $I_E$ contains $N$ regions altogether. The maximal height of a region $A_i$ is computed with function $H(A_i)$ and the maximal width with function $W(A_i)$:

$$SE = \{A_i | (50 \leq W(A_i) \leq 100) \wedge (60 \leq H(A_i) \leq 100) \wedge (A_i \in I_E)\} \; i = 1, 2, \ldots, N \quad (11.1)$$

The formula shows the selected lower limits and upper limits for height and width. Regions are selected whose heights and widths are within the boundaries. If the boundaries in the formula are used for the image $I_E$ that was calculated with morphological operators that were applied to $I_T$, then $SE$ contains only one region that represents the poster $A_i$. The formula reveals that the boundaries can be selected very widely and gives hints as to how reliable the depicted algorithm probable will be. It is expected that such an algorithm is relatively resistant against noise and dynamic changes like alterations of the illumination or the object size in an image. The coordinates of the region are also available. For example, this makes it possible that the region can be cut out for further examinations.

### 11.1.2
### Segmentation with Highpass Filtering

We now show how highpass filtering can be applied to the RGB image $I_O$ that was taken from a Pioneer 1 robot. In this case an algorithm was implemented to detect the second poster in $I_O$. This is the poster that is positioned right beside the neon lamp and opposite to the fire extinguisher. Illumination differences also exist, but are not so pronounced as in the poster that was detected with the algorithm that uses the Gabor filter. The size of the highpass filter mask used was $29 \times 29$. A highpass filter may be appropriate if low frequencies are to be eliminated. The larger the size of the highpass filter mask used the more the frequency domain increases, which may pass into the direction of lower frequency. Therefore, a large highpass filter mask lets pass low frequencies. The highpass filter was applied to the blue channel image $I_B$ that was extracted from the image $I_O$. The result of this convolution can be viewed in Figure 81.

**Figure 81** Highpass filtered image $I_H$

The figure shows that the illumination differences could be defused. The Gabor and highpass filter are both frequency selective. The highpass filter is followed by the dynamic threshold operator that was adjusted with the offset *of* of 30. The resulting binary image is processed with morphological operators. Finally, poster $A_i$ is selected. This occurs by the use of the features height and width for the remaining regions in the image $I_E$. The upper and lower boundaries for these features can be seen in the formula:

$$SE = \{A_i | (40 \leq \mathrm{W}(A_i) \leq 60) \wedge (100 \leq \mathrm{H}(A_i) \leq 200) \wedge (A_i \in \mathrm{I_E})\} \; i = 1, 2, \ldots, N\} \quad . \tag{11.2}$$

If these boundaries are chosen for the image $I_E$, then there remains only one region $A_i$ in $SE$ that represents the desired poster in the RGB image $I_O$, because the zooming of the blue channel was not necessary. The convolution was executed in the spatial domain with a filter mask. The algorithm based on the highpass filtering permits wide ranges for the width and height features of the regions to be selected widely, just as by the algorithm that uses the Gabor filter.

### 11.1.3 Object Selection with a Band Filter

The light switch in the RGB image $I_O$ is to be found. An algorithm is now explained that uses a band filter $I_{BF}^F$ for the object detection. In this case the convolution takes place in the frequency domain. The superscript symbol F denotes that the spatial image $I_{BF}$, which is the band filter, is transformed into the frequency domain. Frequencies that are outside of a frequency band that has the normalized lower boundary $f_{min}$ and normalized upper boundary $f_{max}$ are selected and put into the result of the filtering. Frequency values may be selected for both boundaries that are in the interval [0; 100]. The algorithm uses 40 for $f_{min}$ and 50 for $f_{max}$. Frequencies are selected that are in the intervals [0; 40) and (50; 100]. The algorithm creates the specified band filter to use for the convolution with the Fourier transformed image $I_Z^F$:

$$I_B^F = I_Z^F \cdot I_{BF}^F . \tag{11.3}$$

The convolution occurs in the frequency domain and is marked with the dot. The result is the image $I_B$ that is retransformed into the spatial domain with the inverse Fourier transform. The usual threshold operator (see Section 2.7) is now applied to $I_B$. $T_{min}$ was initialized with 130 and $T_{max}$ with 170. The binary image $I_T$ was processed with morphological operators. Then regions were searched for that were constituted by coherent pixels. The light switch $A_i$ was selected by the use of the features height and width. The used boundaries are shown in the formula:

$$\begin{aligned} SE = \{A_i | (30 \leq W(A_i) \leq 50) \wedge (40 \leq H(A_i) \leq 60) \wedge \\ (A_i \in I_E)\}\ i = 1, 2, \ldots, N \end{aligned} . \tag{11.4}$$

### 11.1.4 Object Detection with the Color Feature

It seems to be favorable to search for the fire extinguisher with its color feature. The red color is very eye catching and singular in the image. The three channels of the RGB image were extracted:

$$I_R(x_A, y_A), \quad I_G(x_A, y_A), \quad I_B(x_A, y_A). \tag{11.5}$$

The gray values in all three channels are defined on the interval [0; 255]. Gray value 0 is used for black and gray value 255 for white. All three channels were utilized for the extinguisher detection. A threshold image was derived from each channel:

$$I_{T_\nu}(x_A, y_A) = \begin{cases} 1 \ \mathit{if} \ T_{\nu\min} \leq I_\nu(x_A, y_A) \leq T_{\nu\max} \\ 0 \ \mathit{else} \qquad \qquad \nu \in \{R, G, B\} \end{cases}. \tag{11.6}$$

$I_{T_\nu}$ is a threshold image that was calculated from channel $\nu$. The used threshold boundaries are now listed:

$$\begin{aligned} T_{R_{\min}} &= 60, \quad T_{R_{\max}} = 130, \ T_{G_{\min}} = 30, \quad T_{G_{\max}} = 65, \\ T_{B_{\min}} &= 30, \quad T_{B_{\max}} = 55. \end{aligned} \tag{11.7}$$

Relatively wide ranges are required. This may be perhaps astonishing, but the image shows color differences within the fire extinguisher. The need for wide ranges makes the use of only one channel for the detection of the fire extinguisher extremely disadvantageous, because too many pixels would remain in the threshold image. Most of these pixels do not belong to the fire extinguisher. Therefore, all three channels are used, and the intersection of the three calculated threshold images is computed:

$$I_{T_{RGB}} = I_{T_R} \cap I_{T_G} \cap I_{T_B}. \tag{11.8}$$

The result of the intersection is a new threshold image $I_{T_{RGB}}$. The value $I_{T_{RGB}}(x_A, y_A)$ only has the value 1 if this holds also for $I_{T_R}(x_A, y_A)$, $I_{T_G}(x_A, y_A)$, and $I_{T_B}(x_A, y_A)$. Otherwise $I_{T_{RGB}}(x_A, y_A)$ obtains the value 0. It is expected that $I_{T_{RGB}}$ contains only a small number of pixels. It is preferred that most of these pixels belong to the fire extinguisher. But Figure 82 shows that this ideal case is not obtained.

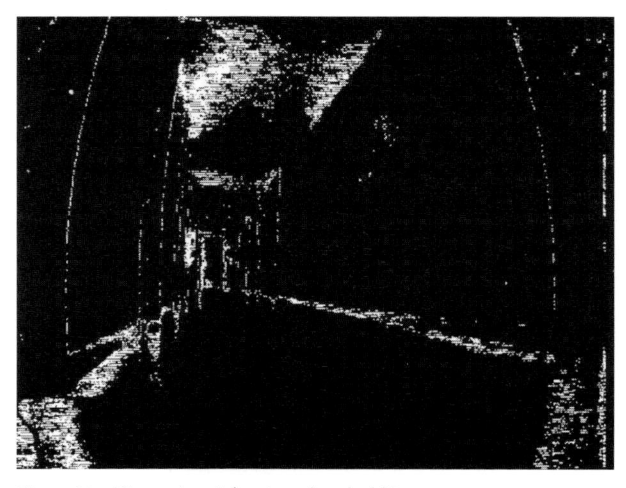

**Figure 82** Fire extinguisher in a threshold image

Many pixels remain in the threshold image $I_{T_{RGB}}$ that do not belong to the fire extinguisher. It can also be observed that not all the pixels that represent the fire extinguisher are selected. This is amazing, because the ranges have been selected relatively widely. The image $I_{T_{RGB}}$ is processed with morphological operators and then the fire extinguisher is selected by the use of the region's features width and height. This strategy is known from the algorithms that have been explained in the previous sections:

$$SE = \{A_i | (15 \leq W(A_i) \leq 20) \wedge (40 \leq H(A_i) \leq 60) \wedge (A_i \in I_E)\} \; i = 1, 2, \ldots, N \quad . \tag{11.9}$$

The unprofitable threshold image $I_{T_{RGB}}$ forces very narrow ranges, which may make it evident that the depicted algorithm seems to be less appropriate in comparison to the algorithms based on Gabor and highpass filtering.

### 11.1.5
### Edge Detection with the Sobel Filter

The Sobel filter is now used to detect the light switch in the image $I_O$. The edge detection happens in the spatial domain. The algorithm uses the mean value filter $I_{MF}$ of size $9 \times 9$ in the first step. The convolution takes place with the image $I_O$:

$$I_M = I_O ** I_{MF} . \tag{11.10}$$

The symbol ** represents the convolution in the spatial domain. Then the edge-detection operator (see Section 10.2.1) was used based on the Sobel filter. The operator was applied to the image $I_M$: the result of the convolution in the spatial domain. The mask of the Sobel filter had a size of $9 \times 9$ and the used calculation specification was 'the amount of the direction difference':

$$I_D(x_A, y_A) = \frac{|\Delta_x I_M(x_A, y_A)| + |\Delta_y I_M(x_A, y_A)|}{2} . \tag{11.11}$$

Subscript D indicates the use of discrete gradients. $\Delta_x I_M(x_A, y_A)$ represents the discrete gradient in the $X_A$ direction and $\Delta_y I_M(x_A, y_A)$ the discrete gradient in the $Y_A$ direction. But $I_D$ is not the image $I_S$ that was gained by the use of a Sobel filter, because this requires an appropriate filter mask. After the edge detection was performed, the edges are approximated with lines, which are included in $I_S$. Then some data is computed in $I_S$, like length, coordinates, and orientation in radians. The applied edge-detection operator enables adjustment of a minimal necessary length. Edges and their approximations respectively are not considered if their lengths fall below the adjusted lower boundary. The image $I_S$ contains the approximated lines. It is a binary image that can be processed with morphological operators. The result of these morphological operations is then decomposed into several regions. Each of these regions represents coherent pixels. Finally, the determination of that region that represents the light switch $A_i$ is carried out. For these purposes region attributes like compactness (see Chapter 2), area, width, and height are used. The following formula shows the used parameterization for the attributes width and height:

$$SE = \{A_i | (20 \leq W(A_i) \leq 30) \wedge (25 \leq H(A_i) \leq 30) \wedge (A_i \in I_E)\} \; i = 1, 2, \ldots, N \quad (11.12)$$

## 11.2 Experiments

In the last sections some algorithms have been elucidated. The experiment is now described that should evaluate the algorithms with regard to their eligibility for indoor exploration projects with autonomous and mobile robots. For these purposes two evaluation criteria were used. One criterion was the range of the interval used for the height attribute. The wider an interval for the height criterion can be selected the better is the valuation of the algorithm. The interval boundaries were determined as wide as possible. This means that the selected boundaries provide exactly one remaining region. This region is the desired region that represents the object to be selected. If the selected lower boundary of the interval were decreased by at least one pixel or the selected upper boundary increased by at least one pixel then this would yield a result with at least two remaining regions. In this case at least one region would be unwanted. In the previous sections sequenced selections were used. For example, regions were first selected that matched to the interval boundaries for the width attribute and then the same strategy was used in conjunction with the height criterion. Also, further attributes are possible like the region area. Such successive selections were not implemented in the experiment, because the comparison of the algorithms would be aggravated. Only the height attribute was considered. Of course, an implementation of an algorithm to be used practically will surely require such successive selections. Further alterations were made to obtain more comparable algorithms. The explained algorithms used the threshold (dynamic and conventional threshold) in most instances. In the experiment the conventional threshold was generally used. Only the algorithm that uses the Sobel operator for the edge detection is implemented without a threshold operation, because the approximated lines were accumulated in a binary image.

The area of the detected region that represented the desired object was the second evaluation criterion. The larger the computed area of the detected region the better was the appraisal of the algorithm. As mentioned before, it is necessary to keep the results of the algorithms as comparable as possible. Therefore, in either case the poster nearby the light switch in the image $I_O$ should be detected from all algorithms in the experiment. The choice of this object happened because of the illumination differences, which can be observed in the region that represents the object. Inhomogeneous illumination is a typical occurrence in indoor exploration projects with autonomous and mobile robots. It was also assured that the examined height intervals could be compared, because in either case a $512 \times 512$ augmentation of the original image was used, also in such cases where a zoomed image was not necessary. The computed values for the two evaluation criteria are listed in Table 2 for all five implemented algorithms.

**Table 2** Evaluation of algorithms.

| Algorithm based on | Max. height interval [pixel] | Area [pixel] |
|---|---|---|
| Sobel filter | [71; 142] | 2483 |
| Color-feature extraction | [66; 140] | 1715 |
| Band filter | [31; 46] | 599 |
| Gabor filter | [45; 511] | 1308 |
| Highpass filter | [98; 156] | 3766 |

It is stated that the area criterion was weighted higher for the evaluation. More information about the object is provided for the reconstruction and recognition if a larger area is detected. But also the range of the height interval must be appraised adequately, because an algorithm that permits a wide interval for the height is less vulnerable to unforeseen disturbances like noise, inhomogeneous illumination, and so on. These aspects yield the conclusion that the algorithm based on the highpass filtering is the best. Sobel and Gabor are considered as equal. With the algorithm based on the Sobel filter, a relatively large object area was gained. The Gabor filter permits a wide interval for the height criterion. The algorithms that used the color-feature extraction and the band filter, did not feature convincing values. These two algorithms are probably not appropriate for indoor exploration if dynamic changes like inhomogeneous illumination occur.

## 11.3 Conclusion

Five algorithms were depicted. The Sobel filter, color-feature extraction, band filter, Gabor filter, or highpass filter were used from the algorithms. Each algorithm used one of these five operators. An experiment was executed. The five algorithms were compared in the experiment with respect to their fitness for indoor exploration projects with autonomous and mobile robots. An image that was taken from a mobile Pioneer 1 robot was used. The image shows the corridor of our research department. The image also shows posters on the wall. One of these posters is characterized by inhomogeneous illumination, which is a typical occurrence in indoor exploration projects with autonomous and mobile robots.

The algorithm that used the highpass filter has proved its robustness. Two evaluation criteria were used for these purposes. One of these two criteria was the calculated area of the segmented region that represented the desired object. The larger the calculated value for the area the better was the evaluation. The other evaluation criterion was the largest range that could be gained for the height interval. This means that the range of the interval could not be increased by at least one pixel without this yielding at least one other object that would be selected in the final result. In that case at least one object would be unwanted.

# 12
# Calibration for Autonomous Video-based Robot Systems

As mentioned before a map should be generated only by a video from an autonomous mobile robot. Therefore, it is necessary that three-dimensional world coordinates $(x^j, y^j, z^j)_W$ with $j = 1, 2, \ldots, n$ of $n$ points in the world coordinate system $\mathbf{X}_W$ are determined. The points may belong to objects and serve the robot for the navigation and the collision avoidance. Therefore, the mounted camera on the Pioneer 1 robot should be calibrated (see Chapter 7).

An approach is depicted following the elucidations in [120]. The robot is equipped with a wide-angle lens. This kind of lens is often used in such projects, but has the drawback that distortions appear in the image because of the fisheye effect. Some further requirements for the explained robot-vision scenario are listed:

1. Production costs must be low for service robots if they are to be profitable.
2. A service robot should execute its tasks in real time.
3. It must be possible to run a robot on different operating systems.

These principles require cheap software. Items two and three are prerequisites for item one. Low productivity would be the result if a service robot were not able to fulfill its tasks in real time. Therefore, fast algorithms must be implemented. Item three will probably increase the gain. The expected higher sales of such a robot allow it to be sold with declining unit costs in comparison to a robot that can be run only on one operating system.

The calibration program SICAST (simple calibration strategy) should help to meet the three elucidated strategies. SICAST is implemented in C++ and is very portable, because it uses only the standard C++ libraries. The development costs for SICAST were very low. Only a C++ compiler and a frame grabber are required. The whole source code is available and was developed according to object-oriented guidelines. These facts support item 2, because algorithms can be replaced by more efficient versions if necessary, and possible bugs in license software will have no affect. A bug in SICAST can be eliminated very rapidly, because the whole source code is accessible. Many existing calibration approaches use license software like MAPLE or MATLAB. Sometimes the calibration procedure is very time consuming.

In the next sections calibration approaches are compared. Algorithms that are implemented in SICAST are depicted, and the object-oriented design of SICAST is

*Robot Vision: Video-based Indoor Exploration with Autonomous and Mobile Robots.* Stefan Florczyk

ISBN: 3-527-40544-5

explained. An experiment was used to check the suitability of the calibration program, and at the end a conclusion can be found.

## 12.1 Camera Calibration for Indoor Exploration

Video-based indoor exploration requires camera calibration. The camera parameters affect the projection of three-dimensional world coordinates $(x^j, y^j, z^j)_W$ of a point $j$ in the world coordinate system $\mathbf{X}_W$ to its two-dimensional image coordinates $(x^j, y^j)_A$ in the image affine coordinate system $\mathbf{X}_A$. If the three-dimensional coordinates of an object are known, they can be collected in a map, which forms the basis for a service robot to fulfill tasks that a watchman or an office messenger has to do.

The calibration of a pinhole camera can be gained with the approach of Faugeras and Mourrain (see Section 7.1.1). In contrast to the other elucidated methods of HALCON and Roger Tsai (see Section 7.1.2), which determine additionally the lens distortion, no initial data is needed.

## 12.2 Simple Calibration with SICAST

### 12.2.1 Requirements

We now elucidate a calibration approach that is based on the proposal of Faugeras and Morrain. A camera, which is mounted on our Pioneer 1 robot, is equipped with a wide-angle lens of low resolution. The images are transmitted to a frame grabber that is integrated in a static computer. The camera's technical data is not available. Therefore, HALCON's operator or Roger Tsai's approach are not suitable for the camera calibration. Another reason for the choice of the proposal of Faugeras and Morrain is the simple process. It is not necessary to take several images from different positions and orientations. MAPLE was used [121] by Faugeras and Morrain in their implementation to execute the mathematical calculations.

The program SICAST needs only the standard C++ libraries. Some algorithms used for the QR-decomposition and singular-value decomposition (SVD) are given in [108]. Also, some algorithms for matrix calculations are taken from [122]. This strategy helps to meet the following listed requirements for the indoor exploration project:

1. Source code is available,
2. Portability is supported,
3. Cheap program development.

In accordance with item three, a three-dimensional calibration object was created. Only a frame grabber and a C++ compiler are necessary for the camera calibration.

Two tins of the same size, which were set one upon the other, constitute the calibration object. The tins were swathed with a sheet of paper that was marked in different heights. This calibration object can be used to measure nondegenerative points. For these purposes the position of the calibration object was changed in every image. The calibration object can be viewed in Figure 83.

**Figure 83** The three-dimensional calibration object

The $x_W^j$ and $y_W^j$ coordinates were varied by the alteration of the calibration object's position within an office cupboard. This should simplify the determination of the coordinates. The exact positioning of the calibration object was supported with a sheet of paper that was printed with standardized squares. For every new positioning a dot was drawn onto the paper. To determine the $x_W^j$ and $y_W^j$ coordinates, only the coordinates of the dot must be measured. The dot also enabled a very proper positioning of the calibration object, because the object could be aligned with the dot. Figure 83 also reveals strong distortions that result from the fisheye effect.

### 12.2.2 Program Architecture

An overview about program architecture is provided in Figure 84, which contains the major classes of the calibration program:

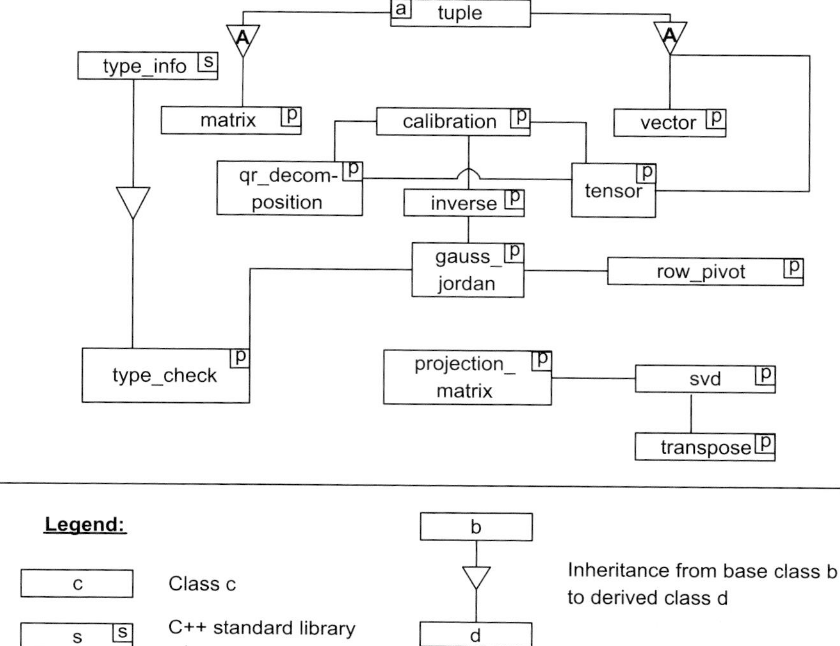


**Figure 84** Program architecture

Classes `matrix` and `vector` are classes on low-level, which are used by many of the shown classes. These associations are omitted in the drawing to preserve the overview. The template concept of C++ was used to implement parameterized classes. This permits the use of different built-in C++ data like `float` or `double` by the same class. The parameterized classes can also use implemented type for fractions. The availability of fractions can be useful for implemented matrix-manipulation algorithms like the Gauss–Jordan method. The class fraction is implemented as template class. Built-in C++ data types can be used from the class fraction. Further utility classes exist among the classes, which are shown in Figure 84. These utility classes are used by the shown classes, but the utility classes are not included in the figure, because the overview would suffer. The figure shows an abstract class `tuple` from which the classes `matrix`, `vector`, and `tensor` are derived. An object of the class `tuple` can not be created. But the derived classes inherit the members of the class `tuple`. These common members can be used from all derived classes. Class `tuple` offers a column attribute to the derived classes and a `show` method that prints out the actual value of the column attribute on the screen. The `show` method is purely virtual and overridden by the derived classes, which possess their own implementations of the `show` method. The size of a `vector` object can be assessed with the column attribute. Class `matrix` provides for the additional dimensioning of a row attribute to register the second dimensions. A `tensor` object is three-dimensional and additively needs attributes for the second and third dimensions.

The projection matrix $D$ is first computed using the class of the same name. The input matrix $I$ with $m$ rows and $n$ columns symbolizes the overdetermined linear equation system. Class `svd` is used to gain the solution of the equation system. Three matrices $M$, $N$, and $Z$ are computed. The matrices $M$, $N$ and the transpose $Z^T$ of $Z$ constitute the result. The size of $M$ is $m \times n$ just as the matrix $I$. The size of $N$ and $Z^T$ is $n \times n$. The equation system

$$I \cdot v_1 = v_2 \tag{12.1}$$

will now be solved. An inhomogeneous solution must be found if $v_2 \neq 0$. The next formula shows the equation system:

$$\begin{pmatrix} \\ M \\ \\ \end{pmatrix} \cdot \begin{pmatrix} \omega_{11}^N & & & \\ & \omega_{22}^N & & \\ & & \cdots & \\ & & \cdots & \\ & & & \omega_{nn}^N \end{pmatrix} \cdot \begin{pmatrix} \\ Z^T \\ \\ \end{pmatrix} = \begin{pmatrix} \\ I \\ \\ \end{pmatrix} . \tag{12.2}$$

A homogeneous solution ($v_2 = 0$) is required for the equation system (12.1). The diagonal elements of the diagonal matrix $N$ are positive or zero. The diagonal elements are denoted by $\omega_{ii}^N$ $(i = 1, 2 \ldots, n)$. If the value $\omega_{ii}^N$ is zero or close to zero, then the corresponding row $i$ of the matrix $Z^T$ constitutes a solution or a proximity solution. Class `svd` uses this fact to determine one (proximity) solution. Other possibly existing solutions must be detected manually. Matrices $N$ and $Z$ must be analyzed for these purposes.

Finally the class `projection_matrix` arrays the vector $v_1$ to the matrix $D$ of size $3 \times 4$. The division of every element in $D$ by the element $c_{34}$ takes place. The result is a normalized matrix $D$ that is needed from class `calibration` as input. The class `calibration` contains algorithms for the formulas (7.5) and (7.8). The matrix $D$ is decomposed into $M$ and $v$. $-M$ is computed. $M$ is then provided for class `qr_decomposition`. The implemented algorithm of the QR-decomposition creates the orthogonal matrix $T$ and the upper triangular matrix $\Psi$ by the decomposition of the matrix $M$. Both matrices are assigned to a `tensor` object that is returned to a `calibration` object.

Finally, the translation vector $\tau$ must be computed. The matrix $-M$ is made available to class `inverse`. A $3 \times 3$ unit matrix is appended at matrix $-M$. The result of the merging is noted with $M'$. $M'$ has six columns. $-M$ is represented by the first three columns of $M'$ and the unit matrix by the last three columns of $M'$.

Class gauss_jordan possesses an algorithm with the same name. The Gauss–Jordan algorithm is used to create a matrix in reduced echelon form [122] from the matrix $M'$. The new matrix is denoted by $M''$, which has the same size as matrix $M'$ $(3\times6)$. The first three columns of $M''$ establish the unit matrix provided that $-M$ is invertible. This condition can be used to prove simply the invertibility of $-M$. The program stops immediately if $-M$ is not invertible and prints an error message on the screen. Otherwise the last three columns of $M''$ are regarded as the inverse of $-M$. The class gauss_jordan is parameterized and permits the use of matrices whose elements can have an allowable domain that is determined by built-in C++ types or elements of the class fraction.

A slightly different algorithm flow is required in the class gauss_jordan if matrices with elements of type fraction are used in comparison to the utilization of floating-point elements. For these purposes the algorithm flow is controlled with if-statements. Class type_check, which is derived from standard C++ class type_info, examines the used elementary type automatically. Otherwise information about the actually utilized type ought to be provided.

## 12.3 Experiments

The three-dimensional calibration introduced was tested with thirty sample points. Indoor exploration is typically accomplished with a camera that is equipped with a wide-angle lens. Such cameras register a large view with one image, but the image is also characterized by strong distortions that hamper the calibration procedure. Such a camera, which was affixed on Pioneer 1 robot, was used in the experiment. None of the measured points were degenerated, because different positions and heights were selected for every point. Therefore, it was found that no value for $x^j_{\mathrm{W}}$, $y^j_{\mathrm{W}}$, $z^j_{\mathrm{W}}$ with $j = 1, 2, \ldots, n = 30$ was measured more than once. The position of the robot and the mounted camera were not changed during the image taking. The three-dimensional world coordinates and the corresponding two-dimensional image coordinates were measured manually. Equation system (7.4) was constructed with the obtained values and solved with SICAST. The gained solution was appraised by the reprojection of the three-dimensional points $(x^j, y^j, z^j)_{\mathrm{W}}$. This was realized with Equation (7.1). The obtained terms $\rho\hat{x}^j_{\mathrm{A}}$ and $\rho\hat{y}^j_{\mathrm{A}}$ were then divided by $\rho$. This enabled the comparison of the computed two-dimensional image coordinates $(\hat{x}^j, \hat{y}^j)_{\mathrm{A}}$ with $j = 1, 2, \ldots, n$, whereby $n$ denotes the number of calculated points, with the empirically measured two-dimensional image coordinates $(x^j, y^j)_{\mathrm{A}}$. The quality of the calibration was evaluated with the mean deviations $x_{\mathrm{mean}}$ and $y_{\mathrm{mean}}$ in $X$ and $Y$ directions:

$$x_{\mathrm{mean}} = \frac{1}{n}\sum_{j=1}^{n}\left|x^j_{\mathrm{A}} - \hat{x}^j_{\mathrm{A}}\right|, \quad y_{\mathrm{mean}} = \frac{1}{n}\sum_{j=1}^{n}\left|y^j_{\mathrm{A}} - \hat{y}^j_{\mathrm{A}}\right|. \tag{12.3}$$

The sample size of $n = 30$ points produced for $x_{\mathrm{mean}}$ the value of 40.9 pixels and for $y_{\mathrm{mean}}$ 33.1 pixels. This result was obviously damaged by five outliers, which were then eliminated. Therefore, another calculation was executed with $n = 25$. This new calculation examined for $x_{\mathrm{mean}}$ the value 20.6 pixels and for $y_{\mathrm{mean}}$ the value 18.8 pixels.

## 12.4 Conclusion

Methods for camera calibration were checked with regard to their suitability for indoor exploration with autonomous and mobile robots. Two approaches were inappropriate due to very time-consuming calibration procedures and the required initial values, which are not available for a camera that is to be calibrated. The last method elucidated does not feature the drawbacks. Nevertheless, its implementation was also useless for the indoor exploration project, because the mathematical calculations were executed with the commercial program library MAPLE. It was claimed that the robot navigation system must be simply portable. This permits only the use of standard libraries. Furthermore, it is demanded that development costs must be as low as possible and the source code must be available. Only an in-house implementation for the camera calibration, which adopts the last approach, can fulfill these requirements.

The program's architecture and functionality was introduced. The calibration program was tested with experiments. The experiments revealed that the utilized wide-angle lens complicated the calibration. First, the result of the experiment was damaged by outliers, which were then eliminated in a subsequent calculation to enhance the capability of the projection matrix. The gained calibration results will be further rectified in the future, because it is expected that a well-calibrated camera helps by the processing of a precise map for navigation tasks. Hence, the calibration program must be modified. Several numerical solutions frequently exist. All these solutions can only be detected manually at the moment. The detection of all numerical solutions should be automated. The best solution should then be selected also automatically from the gained solution set. But further automations are intended.

Outliers should also be detected from the calibration program. The calibration starts with $n$ sample points for which the projection matrix is computed. The quality of the used points can be tested. Therefore, the measured three-dimensional points are reprojected with the computed matrix. The reprojection reveals outliers. The calculated two-dimensional coordinates must only be compared with the acquired two-dimensional empirical data. If the reprojection manifests $m$ outliers, then they must be eliminated from the sample set. The calibration must then be executed again with the new sample set. The automation of this strategy is desired.

Alternative algorithms should be implemented to solve Equation (7.4). The solutions, which will then be created in different ways, should be compared among each other.

# 13
# Redundant Robot-vision Program for CAD Modeling

A new method for CAD (computer-aided design) modeling from image data is introduced. For these purposes the following sections are based on [123], but the explanations are more detailed. In particular, a more extensive experiment is reported. The method was used within an RV (robot-vision) program that was designed for the explained robot-vision scenario in which a mobile robot autonomously creates a map of its office environment. The reconstruction of three-dimensional objects is one important aspect in the map creation. The reconstructed objects should be registered with their three-dimensional world coordinates in the map.

When the robot starts the map generation, it does not possess information with regard to the distances and positions of the objects in the office environment. This is a difficult hurdle for a successful map creation. The new method helps the robot by object finding and reconstructing if information about the actual object distance to the camera is not available.

The dynamic character of the indoor exploration program can aggravate the object detection. The designer of an RV program must recognize the possible disturbances and handle these in the program. In certain cases the effects can damage the object recognition very greatly. For example, large areas of the object to be detected can be occluded in such a way that object recognition is impossible. Occlusions are a challenge for an RV program if they emerge in such a way that the object recognition is not unfeasible. Among the occlusions of object areas other difficulties can appear.

The object parts of the object to be detected can be mixed up with object parts that belong to the occluding object. For example, a chair, whose legs can look very similar to table legs, can occlude a table that is to be detected. An RV program could wrongly consider the chair legs as table legs.

The use of color information is also problematical, because the gray values in the image can vary. This depends on the artificial and natural illumination conditions and the actual object position.

An object that is positioned nearby a window can be difficult to detect because of possible solar irradiation that may overexpose the image. The overexposure can mean that the entire desktop will be represented in an image only by fragments.

Sometimes the object to be detected can be merged with another object that is located very close to the desired object. In this case it can be difficult to separate the objects. If the separation fails, the object detection can be confused, because the

*Robot Vision: Video-based Indoor Exploration with Autonomous and Mobile Robots.* Stefan Florczyk

ISBN: 3-527-40544-5

expected attribute values will differ from the actual measured attribute values. For example, in the case of table detection the attributes height and width can have unforeseen values.

The desired object can also be confused with another object although no occlusions are present. This can occur if the object is similar to the desired object. For example, if a table is to be detected, it can be mistaken for a chair. A chair often also has four legs just as a table. But the number of chair legs is not always four, because the number of chair legs depends on the design. Desktops and chair seats can also be very similar.

Three-dimensional objects often show different views if images are taken from different positions, especially if the camera moves around the object. But all views can not be provided often with the aid of a robot, because some positions are not achievable. A position can be occupied by another object.

The new method, which is now explained, permits the object detection if an object possesses several views. Different distances between the camera and the object can also be handled with the new method. The new method was utilized within a developed RV program to prove the suitability of the method. The RV program was designed in a way that it should be able to handle some of the elucidated unforeseen effects. The new method is depicted in the next section. The design of the method is introduced and then an example is outlined in which an object is modeled with the new method. The developed RV program is explained. For these purposes the program's class design is shown and some specifications are elucidated. The implementation shows the problem handling if unpredictable effects occur. A redundant program was implemented, to enhance the robustness of the object recognition. The introduced design guidelines for redundant robot-vision programs were adhered to. An experiment was executed to prove the eligibility of the RV program and the new method. Sample images that include many of the depicted problems were used in the experiment. A conclusion is given at the end.

## 13.1 New CAD Modeling Method for Robot-vision Applications

### 13.1.1 Functionality

The design of the novel method ICADO (invariant CAD modeling) is introduced in this section. Three-dimensional object reconstruction from image data (see Sections 5.6 and 5.7) will be supported with ICADO. The method utilizes edge models that are conventional graphs for the objects and is therefore oriented at B-rep (see Section 5.2). ICADO is suitable for use within RV programs. The use of the CAD model within the implementation should avoid data-type conversions, which can occur if a CAD database is used. ICADO should be appropriate for object reconstruction if an object can have different distances to a camera. This is the case for the indoor exploration with an autonomous and mobile robot. It may often occur in the scenario that the robot is disadvantageously positioned with regard to the object to be reconstructed. In this case an approximate guess of the object position and distance shall be gained with ICADO. The appraised data can then serve to drive the robot to better positions from which images of more useful quality can be taken. Mostly, three-dimensional objects show several views. A successful recognition with ICADO requires that a specific ICADO model must be generated for every view. ICADO is oriented at B-rep. ICADO does not use absolute measurements for the object modeling, because the values differ in the image if the distances between a camera and an object can change.

To handle these features, ICADO works with norm factors, which are assigned to the edges of the model. Fuzzy knowledge about the expected length of an object part in pixels, which is represented by an edge in the model, must be provided for only one edge. The edge is designated as the reference quantity. All measurements for other edges can be deduced from the reference quantity. Fuzzy knowledge for the reference quantity can be provided with an interval. The domain of the interval that is determined by a lower and upper boundary contains values that probably match to the actually measured length of an object part in pixels that is represented by a reference quantity. The use of an interval allows the measured length to vary within certain boundaries.

The task of the implemented RV program is the detection of that object part that is modeled by a reference quantity. If the particular edge was detected, its length must be determined in pixels. The computed value must suit the domain of the interval that explains the reference quantity. If a detected edge fulfils this condition, then the two-dimensional coordinates of the edge's two endpoints are examined from the image data. These endpoints are modeled in an ICADO model with nodes in the conventional graph. Nodes in an ICADO model generally constitute the endpoints of an edge or connections between edges. A complete ICADO model can now be computed using known norm factors and the determined node coordinates. All unknown edge lengths and node coordinates must be determined. The reference quantity $a_0$ is divided by the edge's norm factor $nf_i$, $i = 1, \ldots, n$, to determine the edge length $a_i$. $n$ is the number of edges that are contained in an ICADO model:

$$a_i = \frac{a_0}{nf_i}, \quad i = 1, \ldots, n. \tag{13.1}$$

A bookshelf was modeled with an ICADO model and can be viewed in Figure 85.

Eleven nodes and 14 edges constitute the bookshelf. Circles are used for nodes whose node numbers are written in the circles. Lengths are attached to the edges. An RV program does not know the actual lengths of object parts in an image, which are represented by edges, because the lengths can differ in images. This holds also for the node coordinates. The top-left corner is represented by an origin $O$. The $X$-axis measures the horizontal distance to the origin and the $Y$-axis the vertical distance in a two-dimensional coordinate system. $x$ and $y$ coordinates belong at least to one edge $r_i$. An RV program must know the values of the norm factors and a permissible domain for the length of a reference quantity. The designer of an ICADO model must examine the object parts that will be represented by the edges of an

ICADO model. One of these object parts must also be selected as the reference quantity. This is the edge zero in Figure 85. How the ICADO model in the figure can be used from an RV program is now elucidated.

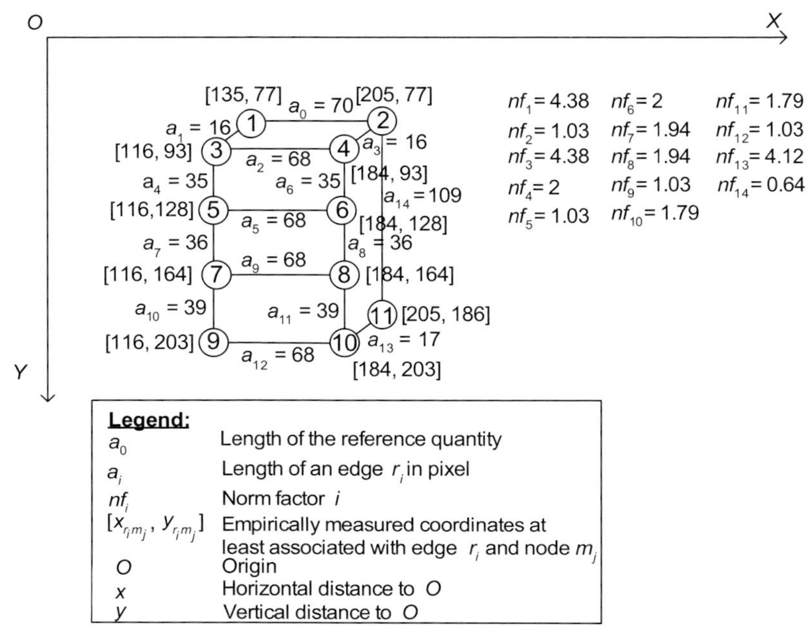


**Figure 85** Bookshelf that is represented with an ICADO model

If the bookshelf is to be reconstructed from an image, an object part must be found that represents the reference quantity. The RV program processes a binary image from the original image. The binary image contains regions of different sizes. Only regions that match to interval [60; 80] for the width and [100; 120] for the height are now selected. The domains, which are determined by lower and upper boundaries, must be ascertained empirically. The boundaries should be selected for intervals as wide as possible to guarantee the detection also if the distance between camera and object fluctuates within certain limits. A segmentation algorithm, which uses the intervals, should detect a region that represents the bookshelf. It is assumed for the following explanations that the discovery of the bookshelf was successful. A rectangle is then created that contains the region. The size of the rectangle is selected as small as possible. The top-left corner of the rectangle can be regarded as the rough estimation of node one that belongs to the ICADO model and the top-right corner as the estimation of node two. The RV program is now able to investigate the length of the reference quantity. For these purposes the value of the $x$ coordinate belonging to node one is subtracted from the value of the $x$ coordinate belonging to node two. Remaining unknown edge lengths and coordinates can now be calculated one after another. For example, the length of edge one can be gained by dividing the length of the reference quantity by the norm factor of edge one:

$$a_1 = \frac{a_0}{nf_1} = 15.98. \tag{13.2}$$

The estimated $\tilde{y}$ coordinate of node three can be obtained by the addition of the length of edge one to the value of the $y$ coordinate of node one. The ICADO model reveals an offset of approximately 19 pixels between node one and node three. The offset must be subtracted from the $x$ coordinate of node one to obtain the valued $\tilde{x}$ coordinate of node three:

$$\tilde{x}_{r_1 m_3} = x_{r_0 m_1} - 19 = 116, \quad \tilde{y}_{r_1 m_3} = y_{r_0 m_1} + a_1 = 92.98. \tag{13.3}$$

The calculation of the two-dimensional coordinates must not consider in either case an offset, as can be verified in Figure 85. The proposed strategy examines a first valuation for the bookshelf. An RV program must contain an algorithm for the classification to decide if the detected region represents a bookshelf in the original image. Further calculations are now executed only on the discovered region to reduce the processing time. Therefore, the region is cut from the entire image. The processed edges in the image clip are compared with the estimated ICADO model. The comparisons in an RV program should permit deviations between the detected edges and the ICADO model, because a too-restrictive strategy will probably eliminate edges that belong to the object to be recognized. The edges that match to the ICADO model probably belong to the bookshelf. An RV program should provide a recognition probability that indicates how confident the detection is. The recognition probability is computed using the selected edges. The more edges selected the higher should be the recognition probability. The computation of the recognition probability happens with an implemented classification algorithm.

It is the task of a designer to determine which object parts of an object to be recognized must be modeled with an ICADO model. Edges within the bookshelf are not part of the ICADO model, because their detection will often be difficult because of shadows within the object, which result from boards belonging to the bookshelf. A designer must recognize such effects during the development of an appropriate ICADO model. If an object possesses different views, it is not possible to guarantee a robust detection if the positions, from which the object will be taken, can change or if the object self-changes its position. In that case a robust detection requires the modeling of several ICADO models. An ICADO model must be created for every possible object view. An RV program that was designed for a scenario where the object's view can not change, can therefore be implemented more simply and should consume less processing time in comparison to an RV program that was developed for a scenario with changing object views.

### 13.1.2
### Program Architecture

Program RICADO (redundant program that uses ICADO) was developed in C++. The program utilizes the commercial image-processing library HALCON/C++. The guidelines for the development of redundant computer-vision programs (see Chapter 10) have been followed. The redundancy of the program is manifested by three methods 'a1', 'a2', and 'a3'. Each of these three methods can compute a result. Therefore, three redundant results can be used. The results are gained using different strategies and parameterizations. Gabor filters are included in all three algorithms, because the direction filtering effect helps to detect a table in an office environment. A desktop is to be found by the examination of horizontal edges and four table legs by the discovery of vertical edges in an image.

The decision for the Gabor filter was influenced by algorithm comparisons (see Chapter 11) for indoor exploration scenarios. Algorithms that were based on the Gabor filter, highpass filter, and Sobel filter gained good results. Therefore, the Gabor filter was used in this work because of its direction filtering effect, as noted before.

A result provides a recognition probability, a ROI (region of interest) that probably represents a table, an edge model of the table, and the size of the discovered ROI. The three methods 'a1', 'a2', and 'a3' are members of class 'algorithms' and are now explained. The drawing in Figure 86 shows the entire class architecture.

The figure does not contain method parameters and attributes to keep the overview. The class 'algorithms' contains a further method with the name 'init'. Method 'init' prepares the use of the other methods in the class and must therefore be called first. The method 'init' converts an RGB image, which is read from the mass storage, into a gray image, because color information is not used for the table detection. The dynamic character of indoor exploration with autonomous and mobile robots often effects different gray values. Therefore, the use of color information will often not be helpful. Zooming to the quadratic size of the gray image is necessary, because the gray image will be converted into the frequency domain. This is executed by the use of the fast Fourier transform whose implementation, realized by a HALCON operator, expects an image of quadratic size. Private members, which belong to the class 'algorithms', store the zoomed and transformed images. These private members can then be used by the other three methods.

The parameterization of the method 'a1' permits the detection of objects if the distance between the camera and the table varies. At the beginning a Gabor filter is generated with a parameterization that permits the discovery of horizontal regions to find the desktop. After the convolution of the transformed gray-value image with the Gabor filter, the result is retransformed into the spatial domain and serves as input for the method 'find_desktop' in class segmentation.

The method will discover the desktop and applies, in a first step, a threshold operator to the input image. Holes with a size between one and ten pixels are then filled in the obtained binary image. Erosion and dilation process the borders. The binary image represents one region that is then decomposed. Coherent pixels in the

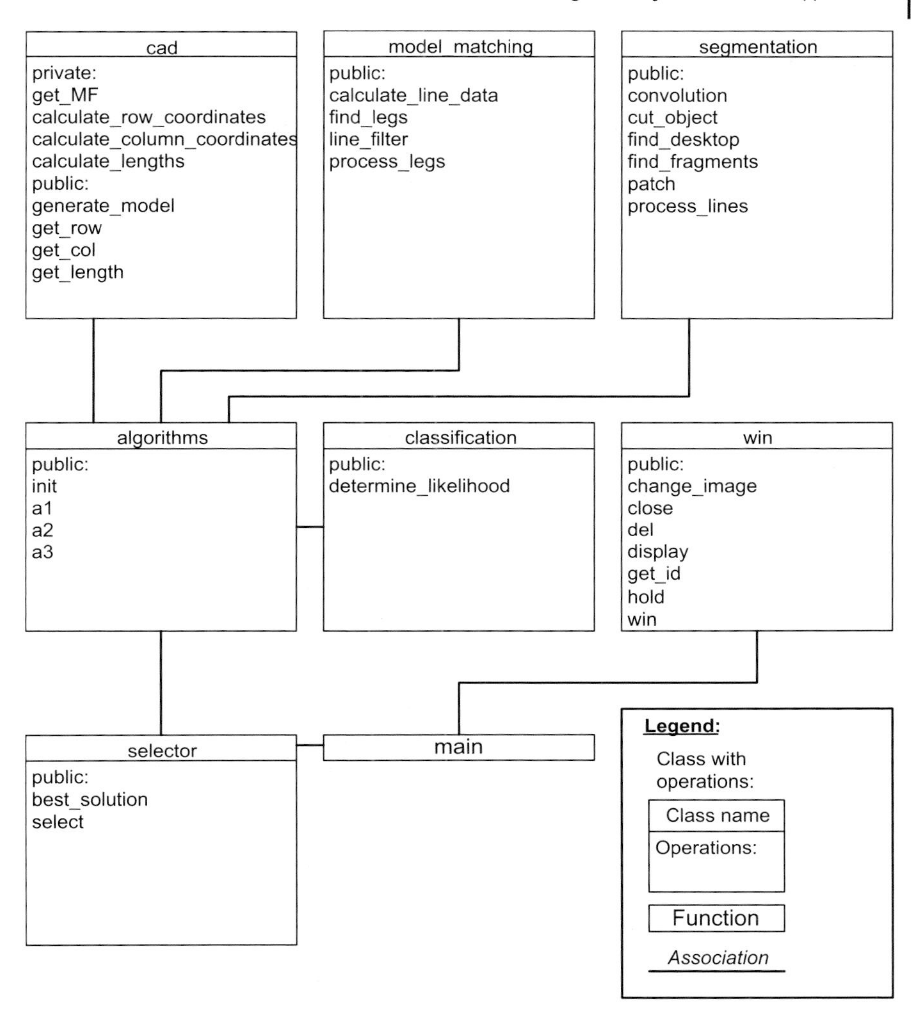


**Figure 86** Class architecture of RICADO

binary image constitute one region. The attributes height and width are used to select regions that match to intervals for both attributes. The interval boundaries for the two intervals are determined by values that are gained from formal parameters belonging to the method 'find_desktop'.

In the next step the number of selected regions is calculated. The desktop detection is considered as successful if only one region can be counted. In this case the recognition probability gets the value 20% assigned. Otherwise it is presumed that probably several fragments establish the entire desktop. The method 'find_desktop' was then unsuccessful and method 'find_fragments' in the class 'segmentation' will support the desktop discovery.

The method 'find_fragments' obtains as input a binary image that is a provisional result of the method 'find_desktop'. The method 'find_fragments' tries to find fragments and if detected, the method assigns the fragments to two variables. One variable produces regions of relatively small width and the second variable regions of larger width, because it is expected that the desktop is broken into smaller and larger regions. A new variable is then used for the unification of the regions. Dilation, which utilizes a circle element with a radius of 3 pixels, is then applied to the unified regions. The dilation should merge desktop fragments that are positioned close together. The recognition probability obtains the value 20 % if now only one region is counted.

Otherwise, method 'patch' tries to find a region that represents a desktop. In this case it is assumed that the region is probably larger than normally expected. A reason for an odd size can be that the table is merged with another object that is positioned in the vicinity of the table. Dilation and erosion are applied to regions that are first selected by the method 'patch'. The disconnection of probably merged objects will occur with erosion that uses a circle with a relatively large radius. Sometimes this strategy affects very small regions that can be eliminated with erosion with a circle that has a radius of two pixels. The method 'patch' proceeds then similar to the method 'find_fragments'. Smaller regions and larger regions are collected separately in two variables. This strategy expects that the former large region is now decomposed into fragments. The unification of the regions that are contained in both variables is assigned to a third variable and then the unified region is processed with morphological operators. The number of regions that are contained in the third variable is computed. If two regions are counted, a line is utilized to merge these regions into one region. Formal parameters provide lower and upper boundaries for an interval that specifies a domain for the attribute width. It is then tested to see if the remaining regions fit to the interval.

After the execution of the selection, the method 'a1' checks the number of remaining regions. Only one region should be counted. The recognition probability is set to 20 % if this is true and then erosion and dilation are applied with circle elements that have in both cases a radius of 2.5 pixels. As mentioned before the detected region is bordered with a smallest rectangle. The calculated top-left corner $[x1_L, y1_L]$ and lower-right corner $[x1_R, y1_R]$ coordinates are used for the processing of an ICADO model. The corner coordinates represent two nodes of the ICADO model. The reference quantity can be computed with these coordinates. Each node that was detected effects an increase of the recognition probability by 10 %. A detected desktop contributes 20 %. Eight nodes in the ICADO model represent the four table legs. Therefore, 100 % can only be gained if a desktop and four table legs are detected.

Method 'generate_model' creates an ICADO model. The method belongs to class 'cad'. The calculated corner coordinates of the smallest rectangle are needed as input from the method. The class 'cad' contains method 'calculate_lengths' to process the length of a reference quantity. The length of the reference quantity is calculated by the difference between the column coordinates of the smallest rectangle's top-left corner and lower-right corner. Method 'get_mf' provides the empirically examined norm factors and is used by the method 'calculate_length'. The height of a table can now be valued as well as all node coordinates of an ICADO model, because the length of a reference quantity and the norm factors are therefore necessary. The examination of the node coordinates occurs with private methods 'calculate_row_-co_ordinates' and 'calculate_column_coordinates', which are used by the method 'generate_model'. The calculated coordinates are then provided to the method 'a1' with the methods 'get_row' and 'get_col'. These two methods expect a node number for which then the belonging row and column coordinate, respectively, is returned. Method 'get_length' makes available the calculated table height $a_1$. The table height is needed from the method 'a1'. A rectangular region is generated with the top-left coordinates of the smallest rectangle and an appraised value for the table height $a_1$. The created rectangular region has top-left coordinates $[x'_{1_L}, y'_{1_L}]$ and lower-right coordinates $[x'_{1_R}, y'_{1_R}]$. It is assumed that the region represents the whole table:

$$[x'_{1_L}, y'_{1_L}] = [x_{1_L} - 20, y_{1_L} - 10], \quad [x'_{1_R}, y'_{1_R}] = [x_{1_R} + 10, y_{1_L} + a_1 + 30]. \tag{13.4}$$

ROI is computed using some constants to ensure that the entire table is contained in the region. An appropriately parameterized Gabor filter is used to discover vertical regions in the ROI. The result is gained in the frequency domain and then retransformed into the spatial domain. The method 'a1' obtains the result and executes a comparison between processed image data and a guessed ICADO model.

The method 'calculate_line_data' of class 'model_matching' is used from the method 'a1' to find with a HALCON operator [4], which is based on the Sobel filter of size $3 \times 3$, vertical edges in a Gabor-filtered image. The detected edges are approximated with lines by the used operator that permits the deviation of a maximal of two pixels between the approximated edge pixels and the line pixels. The operator is also adjusted in such a way that the discovered regions must have at least a length of three pixels. A shorter region is not approximated by a line. The coordinates of the two endpoints belonging to every line are also available.

The number of detected lines is computed with the method 'calculate_line_data'. If no lines are generated, the method aborts immediately. The method 'calculate_line_data' computes for every created line its orientation in radians, the length, and the center coordinates.

The method 'a1' obtains the number of the created lines and then calls the method 'find_legs' that belongs to the class 'model_matching'. The approximate lines are compared with the calculated ICADO model using the calculated line data. The comparison uses relatively relaxed boundaries to permit slight deviations between the ICADO model and the approximate lines. If a line matches to the ICADO model, a region is generated with the thickness of one pixel that has the line's center coordinates and the line's orientation. A larger length is chosen for the region in comparison to the line, because possible disruptions between lines, which represent a table leg, will be patched with this strategy. The comparisons provide a variable to the method 'a1' in which all newly created regions, which probably represent table legs, are accumulated.

The method 'a1' proceeds and calls the method 'process_lines' of the class 'segmentation'. The method 'process_lines' obtains the newly generated variable, which probably contains approximations for table legs, as input. Erosion and dilation are applied to the created regions. A circle element with the radius of three pixels is used in both cases. Holes with sizes between one and ten pixels are filled. Formal parameters determine the boundaries for an interval that defines the allowable domain for the height attribute. Only regions whose sizes match to the interval are now selected. The method 'a1' obtains the selected regions. Regions that are positioned close together will be merged with the use of dilation, because it is assumed that the connected regions probably represent the same table leg.

A classification algorithm now determines the recognition probability. The method 'determine_likelihood' of class 'classification' realizes this. The method compares the node coordinates with the lower and upper areas of the computed regions to examine the recognition probability. Also, these comparisons allow small deviations. A recognized matching effects that the detected node in the ICADO model is labeled as discovered. If all four table legs are properly represented by the created regions, the method 'determine_likelihood' contributes maximal 80 % if all eight nodes are marked as detected. The method 'a1' then obtains the calculated likelihood and adds the value to the recognition probability, which is provided to the calling environment.

Another process is implemented in the method 'a2'. First, the method looks for vertical regions to detect table legs in an image. A Gabor filter is parameterized accordingly to execute a convolution in the frequency domain. After the convolution is performed, the result is retransformed into the spatial domain. The method 'calculate_line_data' is also used from the method 'a2', but with another parameterization. The created approximate lines for probably detected regions are provided to method 'line_filter' that belongs to the class 'model_matching'.

The method 'line_filter' will select only vertical lines from the set of all created lines. To perform this task, the method 'line_filter' is supplied with calculated line data like endpoint coordinates, orientations, and lengths. The proposed design guidelines require that the first calculation attempts will use only fast algorithms. The method 'line_filter' realizes this demand with a flag that indicates as a function of its allocation whether vertical lines with only positive orientations are found or additionally lines with negative orientations. Processing time can be saved if only lines with a positive orientation need be examined. First, the method 'line_filter' is parameterized to find only lines with a positive orientation and creates, for the detected lines, regions with the thickness of one pixel. The method uses the orientations and center coordinates of the corresponding lines for the creation of the regions. The region lengths are larger than the line lengths to effect that possible disruptions between lines that probably represent the same table leg are eliminated. At the end a new variable contains the created regions. Formal parameters provide lower and upper boundaries for an acceptable domain of the height attribute. Those created regions that match to the interval, are selected. The method 'line_filter' decides now which of the generated regions are probably table legs by the comparison of two regions $(A_i, A_j)$. Upper $(e,f)$ row coordinates or lower (0) row coordinates must have values at close quarters or equal values:

$$\text{dev} \geq \left| y_{e\,A_i} - y_{f\,A_j} \right| \vee \text{dev} \geq \left| y_{0\,A_i} - y_{0\,A_j} \right|, \quad i,j = 1, \ldots, n,\ i \neq j. \tag{13.5}$$

An admissible tolerance is denoted with dev. Column coordinates that belong to the regions may be maximally at a distance max and additionally it must hold that the distance between the regions does not fall below the value min:

$$\min \leq \left| x_{0\,A_i} - x_{0\,A_j} \right| \leq \max,\ i,j = 1, \ldots, n,\ i \neq j. \tag{13.6}$$

Formal parameters provide values for the minimally and maximally permitted distances and are ascertained empirically. The measured distances between a camera and a table and the corresponding distances between table legs in images must be acquired. Regions that fulfill the requirements are collected in a new variable and unified into a single region. A rectangle that is as small as possible is then created to border the region. It can be expected that such a rectangle probably covers a region that represents the table legs, but not the desktop. The desktop will also be included. This happens using the top-left and lower-right coordinates of the smallest rectangle. Values are added to these coordinates. The resulting values are used for the creation of a larger rectangular region that also covers the desktop.

Further computations are restricted to this ROI that is cut from the zoomed image. The method 'a2' obtains the calculated ROI and forwards it to method 'convolution' that is parameterized in such a way that the desktop should be found. The earlier detected table legs are not considered from the method 'a2'. The table legs were only discovered to determine the ROI that is first convoluted with a Gabor filter. The desktop will then be found with the method 'find_desktop'.

At best only one region persists. Otherwise additional calculations are required to isolate the desktop. These computations use attributes area, compactness (see Chapter 2), and circularity (see Chapter 2) to select regions that match to intervals that specify the acceptable domain for the attributes. The limits of the intervals are gained from formal parameters provided by the method 'a2'.

Fragments are detected if the method 'a2' can not count any regions after the method 'find_desktop' has finalized. Perhaps an entire region, which represents a desktop, is broken into several parts and therefore a similar strategy as in the method 'a1' is used. Smaller and larger regions are collected separately in two variables. The allowable extensions are defined with two intervals. The regions that are contained in both variables are then unified in a new variable and erosion and dilation are applied. Fragments that are close together are connected with a dilation that uses a relatively large structuring element. Small unwanted regions will be deleted using the erosion. These elucidated operations are applied if required. Only one region should remain. If this is true, the recognition probability is set to 20 %, because it is presumed that this region represents a desktop.

If no region was found, then a desktop was not localized. The method 'a2' was unsuccessful and finalizes immediately. Otherwise a discovered region is processed with erosion and dilation in the method 'a2'. These morphological operators are executed with a circle element that has the radius of 0.5 pixels. The resulting region probably represents a desktop and is bordered with a rectangle that is as small as possible. Data for a reference quantity and an estimated ICADO model can now be gained with rectangle's upper-left $[x_{2_L}, y_{2_L}]$ and lower-right $[x_{2_R}, y_{2_R}]$ corners. The data is calculated with the method 'generate_model' and serves the method 'a2' for the discovery of the table legs and a ROI that will include a table with top-left corner

$$[x'_{2_L}, y'_{2_L}] = [x_{2_L} - 4 \cdot \text{dev}, y_{2_L} - 10 \cdot \text{dev}] \tag{13.7}$$

and lower-right corner

$$[x'_{2_R}, y'_{2_R}] = [x_{2_R} + 4 \cdot \text{dev}, y_{2_R} + 5 \cdot \text{dev}]. \tag{13.8}$$

The boundaries for the ROI are calculated using constant dev to get a relatively wide region with regard to the used corner coordinates. The ROI is calculated with method 'cut_object' also used in the method 'a1'. This also holds for the methods 'convolution', 'calculate_line_data', and 'find_legs'. The strategy, which the method 'find_legs' implements, is then different in comparison to the method 'a1'.

The method 'a2' then uses the method 'process_legs' that belongs to the class 'model_matching'. The class 'model_matching' involves private members, which now contain regions. The contents of the private variables have been calculated before from other members of the class 'model_matching'. These members probably include table legs and are used from the method 'process_legs'. The method processes the regions with erosion and dilation using a circle element with radius 0.5 in both cases. Holes are then filled, provided the area is between one and ten pixels. The height attribute is then used for the selection of regions whose heights match to an interval. The limits of the interval are determined by formal parameters of the method 'process_legs'. The method 'a2' now obtains the remaining regions, which are probably table legs. The recognition probability must now be examined by method 'a2' with the classification algorithm of the method 'determine_likelihood'. The calling environment then obtains the computed recognition probability.

The method 'a3' shows a quite similar strategy as the method 'a2'. Both methods differ especially in the parameterization, but differences in the algorithms can also be observed sometimes. Flags are offered by some methods as formal parameters. The flags enable different algorithm flows to be selected in the methods. The flags must then be parameterized from the methods 'a2' or 'a3' accordingly to choose the desired algorithm flow. The algorithm flows differ especially in the applied sequence of image-processing operators.

The method 'convolution' is used by the method 'a3' and applies an appropriately parameterized Gabor filter to detect vertical regions in the frequency domain. The methods 'calculate_line_data' and 'line_filter' are called from the method 'a3' to find table leg candidates. A rectangle is created with a different size with regard to the generated rectangle in the method 'a2'. The rectangle should cover a table. The next step is the discovery of the desktop with the method 'find_desktop'. This strategy is also quite similar as in the method 'a2', but additional selections follow using the attributes 'compactness' and area size. These selections should yield exactly one region. Otherwise the method 'patch' is called. But the algorithm flow differs slightly in comparison to the method 'a1'. The method 'patch' begins with the choosing of fragments. Two fragment sets are collected. One set inserts regions whose width and area attributes match to intervals that define acceptable domains. A second set includes regions whose heights and areas fit to defined upper and lower limits for both criteria. Both sets are collected in two variables. The content of both variables is then unified in a third variable and processed with erosion that uses a circle element of radius 0.5 pixels. The remaining calculations are the same as if the method 'patch' is called from the method 'a1'. A recognized desktop then effects the adaptation of the recognition probability and table legs are discovered in the computed ROI. Finally, for the method 'a3' is the calculation of the recognition probability that is finally returned.

Method 'select' in class 'selector' controls the calling of the three methods 'a1', 'a2', and 'a3'. The method 'select' prepares the use of these three methods with the call of the 'init' method that is also a member of the class 'algorithms'. First, the method 'a1' provides a result $f_1$ that includes a recognition probability, a calculated area of a ROI, an edge model of a probably detected table, and the ROI's top- and lower-corner coordinates. Arrays are used to store the information. One array of size three is used to store the calculated recognition probabilities of all three methods 'a1', 'a2', and 'a3'. A second array of size three is used to store the ROIs calculated by all three methods and so forth. A second redundant result is necessary to meet the design guidelines. The guidelines demand that a final solution $\mu = f_j, j = 1, 2$, must be computed using at least two results. The method 'select' asks therefore a second redundant result from the method 'a2'. The obtained computations are also stored in the arrays, which already contain the calculations of the method 'a1'. These two results are analyzed from the method 'select' with regard to their eligibility as the final solution. A result is chosen as final solution if it has the recognition probability of at least 50 % and additionally reaches the highest probability. If both results have the same recognition probability of at least 50 %, then the result is chosen that was computed from the method with the lowest number in its name. This is the method 'a1':

$$SE = \{f_j | \varpi_j \equiv \max_i \varpi_i >= 50 \wedge i = 1, 2 \wedge 1 \leq j \leq 2\}, \tag{13.9}$$

$Q$ = The set of all indices that denote an element in $SE$,

$$\mu = \begin{cases} f_j & \textit{if } |SE| = 1 \quad j \in Q \\ f_j & \textit{if } |SE| = 2 \quad j \equiv \min\limits_{i \in Q} i \, j \in Q\,. \\ 0 & \textit{if } |SE| = 0 \end{cases} \tag{13.10}$$

Figure 87 shows data that was calculated from the methods 'a1' and 'a2'.

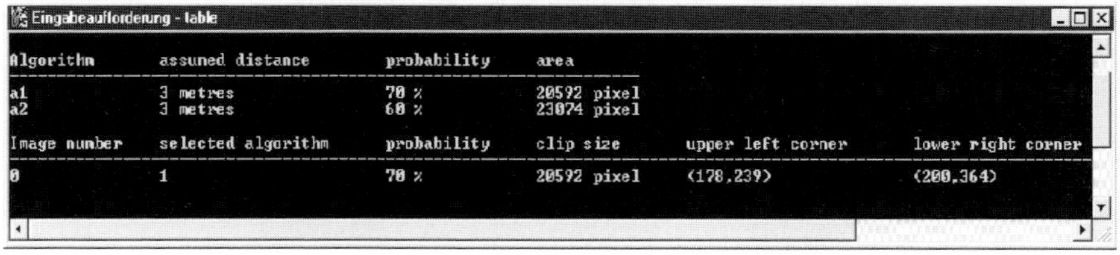

Eingabeaufforderung - table

| Algorithm | assumed distance | probability | area |
|---|---|---|---|
| a1 | 3 metres | 70 % | 20592 pixel |
| a2 | 3 metres | 60 % | 23074 pixel |

| Image number | selected algorithm | probability | clip size | upper left corner | lower right corner |
|---|---|---|---|---|---|
| 0 | 1 | 70 % | 20592 pixel | (178,239) | (200,364) |

**Figure 87** Report for an examined image

The report shows that the analyzed image had the ID zero. The distance of three meters between table and camera was guessed. The confidence of 70 % was gained for the recognition with the method 'a1'. The size $\vartheta_1$ of an image clip was 20 592 pixels. The calculated confidence of the method 'a2' was lower in comparison to the gained confidence of the method 'a1' and reached only 60 %. The calculated size of an image clip $\vartheta_2$ was 23 074 pixels. According to the previous explanations the result of the method 'a1' must be selected as the final result. The result yields the highest recognition probability that is larger than 50 %. For the selected result the coordinates of the top-left corner and the lower-right corner of the image clip (ROI) are printed in the last row of the report.

The method 'a3' must be additionally called from the method 'select' if no result can be selected. In this case the explained conditions could not be fulfilled. The arrays that were used to store the data generated from the methods 'a1' and 'a2' are also used for the data gained from the method 'a3'. Three computed results $f_i$, $i = 1, 2, 3$, are now available. The final solution $\mu = f_j, j \in \{1, 2, 3\}$, will now be chosen from these three results. A final solution can only be selected from these three results if the sum of all three computed recognition probabilities $\varpi_i$ is at least 40 %. Otherwise a final solution $\mu$ can not be yielded. If the sum of all calculated probabilities reaches at least 40 %, then further requirements must be fulfilled before a final solution is selected. A permissible domain is defined for the ROI's size. The domain is defined by an interval with an upper boundary $T_\mathrm{u}$ and a lower boundary $T_\mathrm{l}$. The boundaries were determined empirically. A final solution must include a ROI whose area is within the domain. This condition represents a plausibility check. ROIs with unexpected areas are refused. If more than one result can be chosen, then it is the result selected that features the highest recognition probability. If more than one result fulfills all the explained conditions and have all the same recognition probability that is the highest, then the final solution is yielded from that method ('a1', 'a2', 'a3') with the lowest number in its name:

$$SE_1 = \{f_i | T_\mathrm{l} \leq \vartheta_i \leq T_\mathrm{u} \wedge \sum_{j=1}^{3} \varpi_j \geq 40 \wedge 1 \leq i \leq 3\}, \tag{13.11}$$

$Q_1$ = The set of indices that denote all elements in $SE_1$,

$$SE_2 = \{f_i | \max_{j \in Q_1} \varpi_j \equiv \varpi_i \wedge i \in Q_1\}, \tag{13.12}$$

$Q_2$ = The set of indices that denote all elements in $SE_2$,

$$\mu = \begin{cases} f_j & if\ |SE_2| = 1 \quad j \in Q_2 \\ f_j & if\ |SE_2| = 2 \quad j \equiv \min\limits_{i \in Q_2}\ i\, j \in Q_2 \\ 0 & if\ |SE_2| = 0 \end{cases} . \tag{13.13}$$

The elucidations have shown that the methods possess a large number of formal parameters. This enables their use if the distance between camera and object varies. If the distance changes it is often necessary that the methods are called with another parameterization. The necessary parameterization for a specific distance must be examined empirically. The parameterization for a distance is then provided to the method 'select'. It is the task of the method 'select' to distribute the values to the methods in the class 'algorithms'.

The method 'select' obtains the values from the method 'best_solution' in the class 'selector'. The method obtains several value sets. Every value set was examined empirically. It is appropriate for a particular distance between a camera and an object. The method 'select' is called from the method 'best_solution' with a parameterization that is appropriate to detect a table that has the distance of three meters to the camera. This is the first guess, which happens on the assumption that in an indoor exploration scenario a robot stands in an office door. The robot takes images from the office. A table is probably often located further away from the door in which the robot stands. If the office has a size of $4 \times 4$ meters, and the table is located nearby a wall opposite to the door, then three meters may be a good guess for the first calculation. The discovery of a final solution is indicated with a flag that is provided from the method 'selection'. The method 'best_selection' now calls the method 'select' again if no final solution is found. The call now provides a parameterization to the method 'select' that is appropriate to detect a table with the distance of about two meters to the camera. If this call is also unsuccessful, another parameterization is transferred to the method 'select' that will find tables with the distance of about one meter to the camera. If these calculations fail, then no final solution can be determined.

A discovered final solution is conducted to the 'main' function of RICADO. The final solution provides the approximate distance and position of a table that was probably discovered. The data enables a navigation program to move the robot to positions from which more premium images can be taken. When images can be gained that show several views of a three-dimensional object, the three-dimensional reconstruction can be supported with ICADO models, which were modeled for different object views.

## 13.2 Experiment

An experiment is now depicted. A camera with a wide-angle lens was used. The camera is mounted on a Pioneer 1 robot. A total of 20 sample images were taken in the experiment. The used sample images show many of the described problems. In every sample image a table to be discovered can be viewed. Also, the recognition probability should be examined if the table was successfully detected. The camera had different distances to the table during the image taking. The distances of about one, two, and three meters were chosen. Four sample images can be viewed in Figure 88.

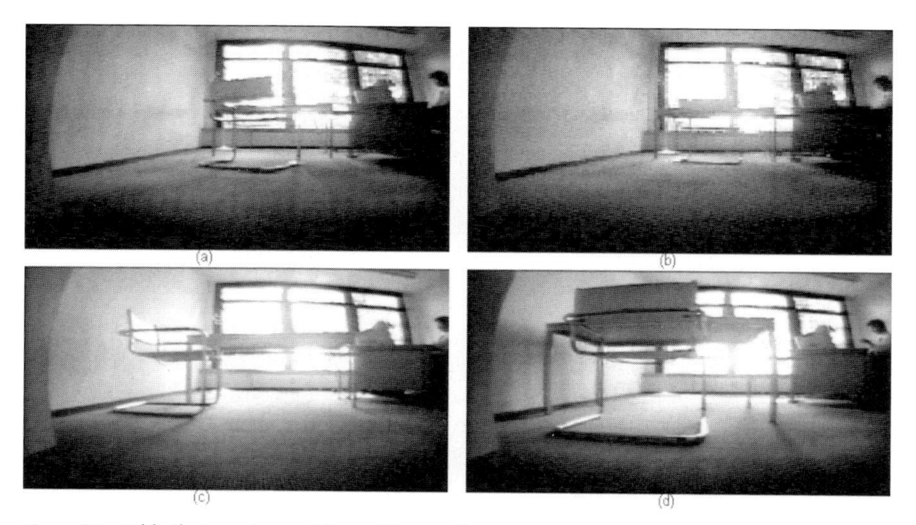


**Figure 88** Table that was imaged from different distances

The image in the top-left corner (a) of the figure shows the table with the distance of about three meters to the camera. A chair is positioned between the camera and table. The camera's view to the table is partially restricted. Some of the chair legs are also located very unfavorably. They are nearby the table legs. A further difficulty is the desk in the right side of the image, which is fixed to the table. The table can be seen very close to a window. This fact effected an overexpose on some table areas in the image. Especially the association between the right-rear table leg and the desktop is broken in the image due to the overexposure. The mix-up of the table with a clip of the widow frame could take place due to the partially similar design of both objects. The table in the top-right corner (b) was also taken from the distance of three meters. Another location was chosen for the chair. A relatively large area of the desktop is occluded. The image in the lower-left corner (c) shows the table that was taken from the distance of two meters to the camera. The table and desk are located directly side by side. A mix-up between chair legs and table legs could occur as the result of the inauspicious placement of the chair. The distance of one meter between the camera and the table was selected as the image in the upper-right corner (d) was taken. Overexposure can be observed in the image area that shows the upper part of the right-rear table leg. The chair is also placed very awkwardly, because it obstructs the camera's view to a large area of the desktop. The table was discovered in all these four sample images from RICADO. Table 3 illustrates the achieved recognition results of RICADO with regard to the 20 sample images whose quality is comparable with the four images shown in Figure 88.

**Table 3** Achieved recognition results of RICADO.

| Likelihood | No recognition | ≤ 40 % | > 40 % and ≤ 60 % | > 60 % and ≤ 90 % |
|---|---|---|---|---|
| Recognition number | 1 | 9 | 7 | 3 |

The test revealed that the discovery of the table only failed in one image. The gained recognition results differ in the other 19 images. A likelihood of less than or equal to 40 % was calculated for nine images. A recognition probability larger than 40 % and less than or equal to 60 % was computed for seven images. The observed likelihood for three images was larger than 60 %, but less than or equal to 90 %. The necessary run-time consumed from RICADO was also measured and printed on the screen after the analysis of the 20 sample images, see Figure 89.

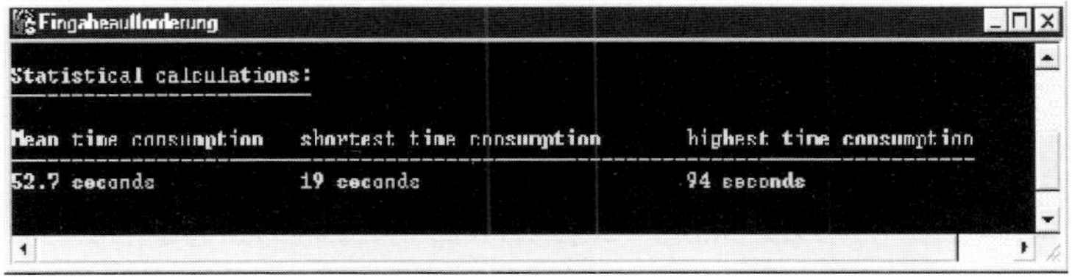

**Figure 89** Performance measurements

RICADO was tested on a computer with a PentiumPro processor with a 200-MHz clock frequency and 128 MB main memory. The shortest measured time consumption was 19 s. This calculation was executed with the two methods 'a1' and 'a2'. The distance between the camera and the table amounted to three meters. RICADO starts the search for the table with the assumption that the table is about three meters from the camera. If the table can not be detected, another search starts with the assumption that the table now has the distance of two meters to the camera. An unsuccessful search would mean that a third search starts. Now the distance of one meter is presumed. Therefore, the highest time consumption of 94 s was observed by the distance of one meter between the camera and the table. 52.7 s were examined for the mean time consumption. Nine images were taken from the distance of three meters, five images from the distance of two meters, and six images from the distance of one meter between the camera and the table.

## 13.3 Conclusion

A new method, ICADO (invariant CAD modeling), was introduced. ICADO supports the creation of CAD models from image data also if distances between the camera and the object can change. ICADO allows the use within an RV (robot-vision) program and avoids data type conversions, which are often necessary if an existing CAD database is used from an RV program. The suitability of ICADO was verified with the program RICADO (redundant program using ICADO). RICADO was programmed for use in a robot-vision scenario.

An autonomous and mobile robot generates a map of its indoor environment. This scenario requires typically the three-dimensional reconstruction of objects whose distances to the robot fluctuate. Some problems can occur in such a scenario due to its dynamic character. The former elucidated design guidelines, which were proposed for redundant computer-vision programs, have been used for the development of RICADO, because it was expected that such a redundant program design can meet problems occurring in the RV scenario.

An experiment was executed, in which 20 sample images, which showed many of the problems, were taken. The images were taken at distances of one, two, and three meters between camera and object. In either case a table should be detected. Some sample images show a chair that obstructs the camera's view. The table is fixed in some images to a desk that can be found nearby the table. The images that were taken at a distance of three meters between the camera and the table show a close placement of the table nearby a window. Some areas of the table are overexposed due to the solar irradiation. The design of some clips of a window frame and the table was similar and could produce confusions. The table was discovered in 19 of the 20 sample images taken. The consumed run-time was measured. A report revealed relatively much processing time that probably results from time-consuming operations in the frequency domain and redundant program design.

The time consumption can probably be diminished if RICADO is ported to a distributed program. The architecture permits porting to be realized simply. But the relatively long run-time seems not really to be a problem for the robot-vision scenario. A completely new map should be generated only once at setup time. This map will only be updated during the running mode. Therefore, a relatively high time consumption is only expected at setup time and not during the running mode.

RICADO provides information about the table's approximate distance and position in an image as soon as the table detection has finished. The information will be used from a robot navigation program to move the robot to more convenient positions. Images of higher quality will be taken from these positions to perform a precise three-dimensional reconstruction. Several views of an object will be modeled in the future with ICADO. A modified version of RICADO must be modeled to test the appropriateness of ICADO for several object views.

# Bibliography

[1] Burgard, W./ Cremers, A. B./ Fox D./ Hähnel, D./ Lakemeyer, G./ Schulz, D./ Steiner, W./ Thrun, S.: Experiences with an interactive museum tour-guide robot, in: Artificial Intelligence, Vol. 114, pp. 3–55, 1999.
[2] Echtle, K.: Fehlertoleranzverfahren, Studienreihe Informatik, Springer, Berlin, 1990.
[3] Wolf, M.: Posturerkennung starrer und nichtstarrer Objekte, Logos, Berlin, 2000.
[4] MVTec: Halcon / C++, MVTec Software GmbH, Munich, 1999.
[5] Koschan, A.: A comparative study on color edge detection, in: 2nd Asian Conference on Computer Vision, (ACCV '95), Volume 3, pp. 574–578, 1995.
[6] Garcia-Campos, R./ Battle, J./ Bischoff, R.: Architecture of an object-based tracking system using colour segmentation, in: 3rd International Workshop in Signal and Image Processing, Manchester, 1996.
[7] Koschan, A.: Minimierung von Interreflexionen in Realen Farbbildern unter Berücksichtigung von Schatten, in: 2nd Workshop Farbbildverarbeitung, Report Nr.1/ 1996 des Zentrums für Bild- und Signalverarbeitung, Technische Universität Illmenau, pp. 65–69, Oktober 1996.
[8] Wiemker, R.: The color constancy problem: an illumination invariant mapping approach, in: International Conference on Computer Analysis of Images and Patterns (CAIP '95), Springer, 1995.
[9] Schuster, R.: Objektverfolgung in Farbbildfolgen, Infix, Sankt Augustin, 1996.
[10] Brammer, K./ Siffling, G.: Kalman-Bucy-Filter: deterministische Beobachtung und stochastische Filterung, 4th edn., Oldenbourg, Munich/ Vienna, 1994.
[11] Kalman, R. E.: A new approach to linear filtering and prediction problems, in: Transactions of the ASME – Journal of Basic Engineering, 82, Series D, 1960.
[12] Denzler, J.: Aktives Sehen zur Echzeitverfolgung, Infix, Sankt Augustin, 1997.
[13] Hansen, M.: Stereosehen: Ein verhaltensbasierter Zugang unter Echtzeitbedingungen, Ph.D. thesis, Report no. 9902, Institut für Informatik und Praktische Mathematik der Christian-Albrechts-Universität Kiel, 1999.
[14] Arnrich, B./ Walter, J.: Lernen der visiomotorischen Kontrolle eines Robotergreifers mit Gaborfiltern, in: Workshop SOAVE '2000 Selbstorganisation von adaptivem Verhalten, Groß, H.-M./ Debes, K./ Böhme, H.-J. (editors), pp. 1–13, VDI GmbH, Düsseldorf, 2000.
[15] Illingworth, J./ Kittler, J.: A survey of the Hough transform, in: Computer Vision, Graphics, and Image Processing, Vol. 44, pp. 87–116, 1988.

*Robot Vision: Video-based Indoor Exploration with Autonomous and Mobile Robots.* Stefan Florczyk

ISBN: 3-527-40544-5

[16] Jones, J. P./ Palmer, L. A.: An evaluation of the two-dimensional Gabor filter model of simple receptive fields in cat striate cortex, in: Journal of Neurophysiology, Vol. 58, No. 6, pp. 1233–1258, 1987.

[17] Abe, S.: Neural Networks and Fuzzy Systems: Theory and Applications, Kluwer, Boston/ London/ Dordrecht, 1997.

[18] Soille, P.: Morphologische Bildverarbeitung: Grundlagen, Methoden, Anwendungen, Springer, Berlin/ Heidelberg/ New York, 1998.

[19] Abmayr, W.: Einführung in die digitale Bildverarbeitung, Teubner, Stuttgart, 1994.

[20] Russ, J. C.: The Image Processing Handbook, 3rd edn., CRC Press, Boca Raton (FL), 1999.

[21] Weszka, J. S.: A survey of threshold selection techniques, in: Computer Graphics and Image Processing, Vol. 7, pp. 259–265, 1978.

[22] Kittler, J./ Illingworth, J./ Föglein, J.: Threshold selection based on a simple image statistic, in: Computer Vision, Graphics, and Image Processing, Vol. 30, pp. 125–147, 1985.

[23] Sahoo, P. K./ Soltani, S./ Wong, A. K. C.: A survey of thresholding techniques, Computer Vision, Graphics, and Image Processing, Vol. 41, pp. 233–260, 1988.

[24] Lee, U./ Chung, S. Y./ Park, R. H.: A comparative performance study of several global thresholding techniques for segmentation, Computer Vision, Graphics, and Image Processing, Vol. 52, pp. 171–190, 1990.

[25] Tani, J./ Fukumura, N.: Learning goal-directed sensory-based navigation of a mobile Robot, in: Neural Networks, Vol. 7, No. 3, pp. 553–563, 1994.

[26] Balakrishnan, K./ Bousquet, O./ Honovar, V.: Spatial learning and localization in animals: A computational model and its implications for mobile robots, Technical report TR# 97–20, Artificial Intelligence Research Group, Department of Computer Science, Iowa State University, 1997.

[27] Berg, de M./ Kreveld, van M./ Overmars, M./ Schwarzkopf, O.: Computational geometry: algorithms and applications, 2nd rev. edn., Springer, Berlin/ Heidelberg/ New York *et al.*, 2000.

[28] Latombe, J.-C.: Robot Motion Planning, 3rd Printing, Kluwer Academic Publishers, 1993.

[29] Brooks, R. A.: A robust layered control system for a mobile robot, in. IEEE Journal of Robotics and Automation, Vol. RA-2, No. 1, pp. 14–23, 1986.

[30] Brooks, R. A.: Intelligence without representation, Artificial Intelligence, Vol. 47, pp. 139–159, 1991.

[31] Freksa, C.: Temporal reasoning based on semi-intervals, Artificial Intelligence, Vol. 54, pp. 199–227, 1992.

[32] Kuipers, B.: Modeling spatial knowledge, in: Cognitive Science, Vol. 2, pp. 129–153, 1978.

[33] Lang, O.: Bildbasierte Roboterregelung mit einer am Greifer montierten Zoomkamera, VDI, Düsseldorf 2000.

[34] Wloka, D. W.: Robotersysteme 1: Technische Grundlagen, Springer, Berlin/ Heidelberg/ New York, 1992.

[35] Paul, R. P.: Robot Manipulators, MIT Press, 7th printing, 1986.

[36] Mohr, R./ Triggs, B.: Projective geometry for image analysis: A tutorial given at ISPRS, http://www.inrialpes.fr/movi/people/Triggs/p/Mohr-isprs96.ps.gz, Vienna, July 1996.

[37] Reece, M./ Harris, K. D.: Memory for places: a navigational model in support of Marr's theory of hippocampal function, in: Hippocampus, Vol. 6, pp. 735–748, 1996.

[38] Kraetzschmar, G./ Sablatnög, S./ Enderle, S./ Utz, H./ Steffen, S./ Palm, G.: Integration of multiple representation and navigation concepts on autonomous

mobile robots, in: Workshop SOAVE '2000 Selbstorganisation von adaptivem Verhalten, Groß, H.-M./ Debes, K./ Böhme, H.-J. (editors), pp. 1–13, VDI GmbH, Düsseldorf, 2000.

[39] Buhmann, J./ Burgard, W./ Cremers, A. B./ Fox, D./ Hofmann, T./ Schneider, F. E./ Strikos, J./ Thrun, S.: The mobile robot RHINO, in: AI Magazine, Vol. 16, No. 2, pp. 31–38, 1995.

[40] Thrun, S./ Bücken, A.: Learning maps for indoor mobile robot navigation, Technical Report CMU-CS-96–121, Carnegie Mellon University, School of Computer Science, Pittsburgh, PA 15213, April 1996.

[41] Russell, S. J./ Norvig, P.: Artificial Intelligence: A Modern Approach, Prentice-Hall, Upper Saddle River (New Jersey), 1995.

[42] Schölkopf, B./ Mallot, H. A.: View-based cognitive mapping and path planning, in: Adaptive Behavior, Vol. 3, No. 3, pp. 311–348, 1995.

[43] Franz, M. O./ Schölkopf, B./ Mallot, H. A./ Bülthoff, H. H.: Where did I take that snapshot? Scene-based homing by image matching, Biological Cybernetics, Vol. 79, Springer, pp. 191–202, 1998.

[44] Franz, M. O./ Schölkopf, B./ Mallot, H. A./ Bülthoff, H. H.: Learning view graphs for robot navigation, Autonomous Robots, Vol. 5, Kluwer Academic Publishers, pp. 111–125, 1998.

[45] Tani, J.: Model-based learning for mobile robot navigation from the dynamical Systems Perspective, IEEE Transactions on Systems, Man, and Cybernetics-Part B: Cybernetics, Vol. 26, No. 3, pp. 421–436, June 1996.

[46] Täubig, H./ Heinze, A.: Weiterentwicklung eines Graphen-basierten Ansatzes als Rahmensystem für eine lokale Navigation, in: Workshop SOAVE '2000 Selbstorganisation von adaptivem Verhalten, Groß, H.-M./ Debes, K./ Böhme, H.-J. (editors), pp. 47–56, VDI GmbH, Düsseldorf, 2000.

[47] König, A./ Key, J./ Gross, H.-M.: Visuell basierte Monte-Carlo Lokalisation für mobile Roboter mit omnidirektionalen Kameras, in: Workshop SOAVE '2000 Selbstorganisation von adaptivem Verhalten, Groß, H.-M./ Debes, K./ Böhme, H.-J. (editors), pp. 31–38, VDI GmbH, Düsseldorf, 2000.

[48] Thrun, S./ Burgard, W./ Fox, D.: A probabilistic approach to concurrent mapping and localization for mobile robots, Machine Learning, Vol. 31, pp. 29–53, 1998.

[49] Elsen, I.: Ansichtenbasierte 3D-Objekterkennung mit erweiterten Selbstorganisierenden Merkmalskarten, VDI GmbH, Düsseldorf, 2000.

[50] Hubel, D. H./ Wiesel, T. N.: Brain Mechanisms of Vision, in: Neuro-vision Systems: Principles and Applications, Gupta, M. M./ Knopf, G. K. (editors), IEEE Press, New York, Reprint, pp. 163–176, 1993.

[51] Büker, U./ Drüe, S./ Hartmann, G.: Ein neuronales Bilderkennungssystem für Robotikanwendungen, in: at – Automatisierungstechnik, Vol. 45, pp. 501–506, 1997.

[52] Capurro, C./ Panerai, F./ Sandini, G.: Dynamic vergence using log-polar images, in: International Journal of Computer Vision, Vol. 24, No. 1, pp. 79–94, 1997.

[53] Bradski, G./ Grossberg, S.: Fast-learning VIEWNET architectures for recognizing three-dimensional objects from multiple two-dimensional views, in: Neural Networks, Vol. 8, No. 7/8, pp. 1053–1080, 1995.

[54] Abbott, L. A.: A survey of selective fixation control for machine vision, in: IEEE Control Systems, Vol. 12, No. 4, pp. 25–31, 1992.

[55] Burt, P. J.: Smart sensing within a pyramid vision machine, in: Proceedings of the IEEE, Vol. 76, No. 8, pp. 1006–1015, 1988.

[56] Califano, A./ Kjeldsen, R./ Bolle, R. M.: Data and model driven foveation, in: Proceedings of the 10th International Conference on Pattern Recognition, pp. 1–7, 1990, cited according to [54].

[57] Krotkov, E. P.: Exploratory visual sensing for determining spatial layout with an agile stereo camera system, Ph.D. thesis, Univ. Pennsylvania, 1987 cited according to [54].

[58] Abbott, L. A./ Ahuja, N.: Surface reconstruction by dynamic integration of focus, camera vergence, and stereo, in: Proc. Second Int. Conf. Computer Vision, pp. 532–543, 1988 cited according to [54].

[59] Marr, D.: Vision, W. H. Freeman, San Francisco, 1982 cited according to [60].

[60] Ahrns, I.: Ortsvariantes aktives Sehen für die partielle Tiefenrekonstruktion: Ein System zur visuell gestützten Manipulatorsteuerung auf der Basis eines biologisch motivierten Sehkonzepts, VDI GmbH, Düsseldorf, 2000.

[61] Horn, B. K. P.: Robot Vision, McGraw Hill Book Company, New York, 1991 cited according to [60].

[62] Rottensteiner, F./ Paar, G./ Pölzleitner, W.: Fundamentals, in: Digital Image Analysis: Selected Techniques and Applications, Kropatsch, W. G./ Bischof, H. (editors), Springer, New York/ Berlin/ Heidelberg, pp. 373–390, 2001.

[63] Sonka, M./ Hlavac, V./ Boyle, R.: Image processing, analysis, and machine vision, 2nd edn., Pacific Grove/ Albany/ Bonn, PWS Publishing, 1999.

[64] Mohr, R.: Projective geometry and computer vision, in: Handbook of Pattern Recognition and Computer Vision, Chen, C. H./ Pau, L. F./ Wang, P. S. P. (editors), World Scientific Publishing Company, Singapore/ New Jersey/ London, pp. 369–393, 1993.

[65] Tönjes, R.: Wissensbasierte Interpretation und 3D-Rekonstruktion von Landschaftsszenen aus Luftbildern, VDI GmbH, Düsseldorf, 1999.

[66] Yakimovsky, Y./ Cunningham, R.: A system for extracting three-dimensional measurements from a stereo pair of TV Cameras, in: Computer Graphics and Image Processing, Vol. 7, pp. 195–210, 1978.

[67] Bao, Z.: Rechnerunterstützte Kollisionsprüfung auf der Basis eines B-rep Polytree CSG-Hybridmodells in einem integrierten CAD CAM-System, VDI GmbH, Düsseldorf, 2000.

[68] Samet, H./ Webber, R. E.: Hierarchical data structures and algorithms for computer graphics, Part II: applications, in: IEEE Computer Graphics and Applications, Vol. 8, No. 4, pp. 59–75, 1988.

[69] Meagher, D.: Geometric modeling using octree encoding, in: Computer Graphics and Image Processing, Vol. 19, pp. 129–147, 1982.

[70] Jackins, C. L./ Tanimoto, S. L.: Oct-trees and their use in representing three-dimensional objects, in: Computer Graphics and Image Processing, Vol. 14, pp. 249–270, 1980.

[71] Brunet, P./ Navazo, I.: Geometric modelling using exact octree representation of polyhedral objects, in: EUROGRAPHICS '85, Vandoni, C. E: (editor), Elsevier Science Publishers, North-Holland, pp. 159–169, 1985.

[72] Carlbom, I./ Chakravarty, I./ Vanderschel, D.: A hierarchical data structure for representing the spatial decomposition of 3-D Objects, in: IEEE Computer Graphics and applications, Vol. 5, No. 4, pp. 24–31, 1985.

[73] Samet, H./ Webber, R. E.: Hierarchical data structures and algorithms for computer graphics, Part I: fundamentals, in: IEEE Computer Graphics and applications, Vol. 8, No. 3, pp. 48–68, 1988.

[74] Jense, G. J.: Voxel-based methods for CAD, in: Computer-Aided Design, Vol. 21, No. 8, pp. 528–533, 1989.

[75] Schmidt, W.: Grafikunterstützes Simulationssystem für komplexe Bearbeitungsvorgänge in numerischen Steuerungen, Springer, Berlin/ Heidelberg/ New York, 1988.

[76] Pritschow, G./ Ioannides, M./ Steffen, M.: Modellierverfahren für die 3D-Simulation von NC-Bearbeitungen, in: VDI-Z, Vol. 135, No.6, pp. 47–52, 1993.

[77] Hook, van T.: Real-time shaded NC milling display, in: Computer Graphics, Vol. 20, No. 4, pp. 15–20, August 1986.

[78] Tamminen, M./ Samet, H.: Efficient octree conversion by connectivity labeling, in: Computer Graphics, Vol. 18, No. 3, pp. 43–51, July 1984.

[79] Kela, A.: Hierarchical octree approximations for boundary representation-based geometric models, in: Computer-aided Design, Vol. 21, No. 6, pp. 355–362, 1989.

[80] Kunii, T. L./ Satoh, T./ Yamaguchi, K.: Generation of topological boundary representations from octree encoding, in: IEEE Computer Graphics and applications, Vol. 5, No. 3, pp. 29–38, 1985.

[81] Vossler, D. L.: Sweep-to-CSG conversion using pattern recognition techniques, in: IEEE Computer Graphics and Applications, Vol. 5, No. 8, pp. 61–68, 1985.

[82] Seiler, W.: Technische Modellierungs- und Kommunikationsverfahren für das Konzipieren und Gestalten auf der Basis der Modell-Integration, VDI GmbH, Düsseldorf 1985.

[83] Brooks, R. A.: Symbolic reasoning among 3-D models and 2-D images, in: Artificial Intelligence, Vol. 17, pp. 285–348, 1981.

[84] Flynn, P. J./ Jain, A. K.: BONSAI: 3-D object recognition using constrained search, in: IEEE Transactions on Pattern Analysis and Machine Intelligence, Vol. 13, No. 10, pp. 1066–1075, 1991.

[85] Glauser, T./ Bunke, H.: Generierung von Entscheidungsbäumen aus CAD-Modellen für Erkennungsaufgaben, in: Mustererkennung: 11. DAGM-Symposium, Proceedings, Burkhardt, H./ Höhne, K. H./ Neumann, B. (editors), Springer, Berlin/ Heidelberg/ New York, pp. 334–340, 1989.

[86] Flynn, P. J./ Jain, A. K.: 3D object recognition using invariant feature indexing of interpretation tables, in: CVGIP: Image Understanding, Vol. 55, No. 2, pp. 119–129, 1992.

[87] Blake, A./ Zisserman, A.: Visual Reconstruction, MIT Press, Cambridge (Massachusetts)/ London, 1987.

[88] Grimson, W./ Eric, L./ Pavlidis, T.: Discontinuity detection for visual surface reconstruction, in: Computer Vision, Graphics, and Image Processing, Vol. 30, pp. 316–330, 1985.

[89] Geman, S./ Geman, D.: Stochastic relaxation, Gibbs distributions, and the Bayesian restoration of images, in: IEEE Transactions on Pattern Analysis and Machine Intelligence, Vol. PAMI-6, No. 6, pp. 721–741, 1984.

[90] Shapiro, L. G./ Lu, H.: Accumulator-based inexact matching using relational summaries, in: Machine Vision and Applications, Vol. 3, pp. 143–158, 1990.

[91] Horaud, P./ Bolles, R. C.: 3DPO: A system for matching 3-D objects in range data, in: From pixels to predicates: recent advances in computational and robotic vision, Pentland, A. P. (editor), Ablex Publishing Corporation, Norwooden (New Jersey), pp. 359–370, 1986.

[92] Wang, W./ Iyengar, S. S.: Efficient data structures for model-based 3-D object recognition and localization from range images, in: IEEE Transactions on Pattern Analysis and Machine Intelligence, Vol. 14, No. 10, pp. 1035–1045,1992.

[93] Otterbach, R.: Robuste 3D-Objekterkennung und Lagebestimmung durch Auswertung von 2D-Bildfolgen, VDI, Düsseldorf 1995.

[94] Zahn, C. T./ Roskies, R. Z.: Fourier descriptors for plane closed curves, IEEE Transactions on Computers, Vol. C-21, No. 3, March 1972.

[95] Winkelbach, S./ Wahl, F. M.: Shape from 2D edge gradients, in: Pattern recognition, Radig, B./ Florczyk, S. (editors), 23rd DAGM Symposium, Munich, Germany, Proceedings, Springer, Berlin, pp. 377–384, 2001.

[96] Brunn, A./ Gülch, E./ Lang, F./ Förstner, W.: A multi-layer strategy for 3D building acquisition, in: Mapping buildings, roads and other man-made structures

from images, Leberl, F./ Kalliany, R./ Gruber, M. (editors), Proceedings of the IAPR TC-7, Oldenbourg, Vienna/ Munich, pp. 11–37, 1997.

[97] Pearson, J./ Olson, J.: Extracting buildings quickly using radiometric models, in: Mapping buildings, roads and other man-made structures from images, Leberl, F./ Kalliany, R./ Gruber, M. (editors), Proceedings of the IAPR TC-7, Oldenbourg, Vienna/ Munich, pp. 205–211, 1997.

[98] Stilla, U./ Michaelsen, E./ Lütjen, K.: Automatic extraction of buildings from aerial Images, in: Mapping buildings, roads and other man-made structures from images, Leberl, F./ Kalliany, R./ Gruber, M. (editors), Proceedings of the IAPR TC-7, Oldenbourg, Vienna/ Munich, pp. 229–244, 1997.

[99] Faugeras, O. D.: Three-dimensional Computer Vision: A Geometric Viewpoint, MIT Press, Cambridge (Massachusetts), London, 1993.

[100] Hartley, R. I.: Estimation of relative camera positions for uncalibrated cameras, in: Computer Vision – ECCV '92, Second European Conference on Computer Vision, Santa Margherita Ligure, Italy, Proceedings, Sandini, G. (editor), pp. 579–587, Springer, 1992.

[101] Faugeras, O. D./ Luong, Q. T./ Maybank, S. J.: Camera self-calibration: theory and experiments, in: Computer Vision – ECCV '92, Second European Conference on Computer Vision, Santa Margherita Ligure, Italy, Proceedings, Sandini, G. (editor), pp. 321–334, Springer, 1992.

[102] Ayache, N./ Hansen, C.: Rectification of images for binocular and trinocular stereovision, in: $9^{th}$ International Conference on Pattern Recognition, Rome, IEEE, Los Alamitos (CA), pp. 11–16, 1988.

[103] Klette, R./ Koschan, A./ Schlüns, K.: Computer Vision: Räumliche Information aus digitalen Bildern, Vieweg, Wiesbaden, 1996.

[104] Echigo, T.: A camera calibration technique using three sets of parallel lines, in: Machine Vision and Applications, Vol. 3, pp. 159–167, 1990.

[105] Wang, L.-L./ Tsai, W.-H.: Computing camera parameters using vanishing-line information from a rectangular parallelepiped, in: Machine Vision and Applications, Vol. 3, pp. 129–141, 1990.

[106] Martins, H. A./ Birk, J. R./ Kelley, R. B.: Camera models based on data from two calibration planes, in: Computer Graphics and Image Processing, Vol. 17, pp. 173–180, 1981.

[107] Izaguirre, A./ Pu, P./ Summers, J.: A new development in camera calibration: calibrating a pair of mobile cameras, in: The International Journal of Robotics Research, Vol. 6, pp. 104–116, 1987.

[108] Press, W. H./ Teukolsky, S. A./ Vetterling, W. T./ Flannery, B. P.: Numerical Recipes in C: The Art of Scientific Computing, 2. edn., Cambridge Univ. Press, Cambridge, 1992.

[109] Tsai, R.: A versatile camera calibration technique for high-accuracy 3D machine vision metrology using off-the-shelf TV cameras and lenses, in: IEEE Journal of Robotics and Automation, Vol. 3, No. 4, pp. 323–344, 1987.

[110] Shapiro, L. G./ Stockman, G. C.: Computer Vision, Prentice Hall, Upper Saddle River (New Jersey) 2001.

[111] Lin, D./ Lauber, A.: Calibrating a camera by robot arm motion, Proceedings ISMCR'95, S3 – Measurement, Performance, Evaluation and Improvement cited according to [33].

[112] Wei, G.-Q./ Arbter, K./ Hirzinger, G.: Active self-calibration of robotic eyes and hand-eye relationships with model identification, in: IEEE Transactions on Robotics and Automation, Vol. 14, No. 1, pp. 158–166, 1998.

[113] Wünstel, M./ Polani, D./ Uthmann, T./ Perl, J.: Behavior classification with Self-Organizing Maps, in: Workshop SOAVE '2000 Selbstorganisation von adaptivem

Verhalten, Groß, H.-M./ Debes, K./ Böhme, H.-J. (editors), pp. 77–85, VDI GmbH, Düsseldorf, 2000.

[114] Böhm, M.: Formulartyperkennung und Schriftaufbereitung in der optischen automatischen Archivierung. Ph.D. thesis, Technical University Berlin, 1997.

[115] Florczyk, S.: A redundant program to handle inhomogeneous illumination and changing camera positions in a robot vision scenario, in: Pattern Recognition: Methods and Applications, Murshed, N. (editor), Proc. VI Iber-American Symp. Pattern Recognition, IAPR-TC3, Florianópolis, Brazil, 2001, pp. 100–105.

[116] Blum, S.: OSCAR – Eine Systemarchitektur für den autonomen mobilen Roboter MARVIN, in: Autonome Mobile Systeme, Informatik Aktuell, pp. 218–230, Springer, 2000.

[117] Florczyk, S.: Evaluation of object segmentation algorithms as components of autonomous video based robot systems, in: Second International Conference on Images and Graphics, Wei, S. (editor), Vol. 4875, SPIE, 2002, pp. 386–393.

[118] Ritter, G. X./ Wilson, J. N.: Computer vision algorithms in image algebra, CRC Press, Boca Raton/ New York/ Washington, *et al.*, 2000.

[119] Stroustrup, B.: Die C++ Programmiersprache, 2. edn., Addison-Wesley, Bonn, Germany, 1992.

[120] Florczyk, S.: A calibration program for autonomous video based robot navigation systems, in: Krasnoproshin, V./ Ablameyko, S./ Soldek, J. (editors), Pattern Recognition and Information Processing: Proceedings of the Seventh International Conference, Vol. I, Minsk: United Institute of Informatics Problems of National Academy of Sciences of Belarus, pp. 129–133, 2003.

[121] Faugeras, O./ Mourrain, B.: On the geometry and algebra of the point and line correspondences between *N* images, Technical Report 2665, Institut National de Recherche en Informatique et en Automatique, Sophia-Antipolis, France, 1995.

[122] Janiszczak, I./ Knörr, R./ Michler, G. O.: Lineare Algebra für Wirtschaftsinformatiker: Ein algorithmen-orientiertes Lehrbuch mit Lernsoftware, Vieweg, Braunschweig/ Wiesbaden, Germany, 1992.

[123] Florczyk, S.: ICADO – A method for video based CAD modeling, in: Krasnoproshin, V./ Ablameyko, S./ Soldek, J. (editors), Pattern Recognition and Information Processing: Proceedings of the Seventh International Conference, Vol. II, Szczecin: Publishing House and Printing House of Technical University of Szczecin Faculty of Computer Science and Information Technology, pp. 321–325, 2003.

# Index



*Robot Vision: Video-based Indoor Exploration with Autonomous and Mobile Robots.* Stefan Florczyk

ISBN: 3-527-40544-5